Astronomy Made Simple

Michael Hamburg

Based on the original edition by Meir H. Degani

Edited and prepared for publication by
The Stonesong Press, Inc.

A Made Simple Book

DOUBLEDAY New York London Toronto Sydney Auckland

Edited and prepared for publication by The Stonesong Press, Inc.
Executive Editor: Sheree Bykofsky
Series Editor: Sarah Gold
Editor: Patricia Barnes-Svarney

A MADE SIMPLE BOOK
PUBLISHED BY DOUBLEDAY
a division of Bantam Doubleday Dell Publishing Group, Inc.
1540 Broadway, New York, New York 10036

Printed in the United States of America
February 1993
Fourth Edition

Library of Congress Cataloging-in-Publication Data

Hamburg, Michael.
Astronomy made simple / by Michael Hamburg; based on the original edition by Meir H. Degani.—4th ed.
p. cm.
"A Made simple book."
Includes index.
1. Astronomy. I. Degani, Meir H. Astronomy made simple.
II. Title.
QB45.H28 1993
520—dc20

92–3726
CIP

ISBN 0–385–26582–4

11 10 9 8 7 6 5

CONTENTS

INTRODUCTION

For as long as people have been conscious of themselves and the external universe, they have regarded the sky with awe and wonder. This fascination with the sky led people to study it systematically, discern regular patterns, and attempt to develop predictions based on these patterns. Astronomy has grown out of our restless minds continually asking questions and seeking answers about how our universe works.

Astronomy is the science of the positions, motions, compositions, histories, interactions, and destinies of celestial bodies. In the course of its development as a science, many of the basic physical laws governing those bodies have become well understood. But not everything has been discovered. It is the nature of scientific investigation to continue observing and testing and to keep looking for the unexpected.

Astronomy is one of the physical sciences, as are physics, geology, and chemistry. The science of astronomy often follows the same procedures as the other sciences. Astronomers construct models that explain the behavior of objects in the sky and describe their origin and evolution. Models begin as educated guesses or hypotheses based on relevant information. When one or more hypotheses appears to be consistent with a model, it is called a theory. Theories are always subject to testing and validation.

A Brief History

Astronomy, one of the earliest sciences, developed as various civilizations learned to associate the appearance and movements of objects in the sky with natural changes on the Earth, such as the cycle of the changing seasons, the change in the length of day and night, and the rise and fall of the oceans. Before our sky was flooded with artificial light that obscures even the brightest stars, people depended on the stars and the sky for all kinds of information. Sailors and overland traders used the sky as a map. Reading the evening sky was one of the first ways people tried to predict the weather. The passage of time was measured by the movement of the Sun and by the phases of the Moon.

Technology now provides us with digital clocks for measuring accurate time, detailed road maps for safe and certain routes, and radar and other such electronic wizardry for guiding us in all types of weather. We no longer need to navigate by the stars and their constellations. We know about our planet without any reference to the other objects in the sky because we now have calendars and the accumulated knowledge of thousands of years of astronomical studies. Men have walked on the moon, robot spacecraft have toured the solar system, and future manned and unmanned missions into space are being planned. Our horizons are now cosmic rather than just terrestrial. Today's astronomy seeks the mysteries of the beginning and end of the universe—once solely the concern of philosophy and religion.

The science itself has also changed. Four hundred years ago, astrology was once part of astronomy. Astronomers of the time cast horoscopes for the rich and powerful. Two major discoveries decreased the emphasis on astrology: The Sun, rather than the Earth, was proven to be at the center of the Solar System; and the existence of a universal force of gravity accurately explained how the planets and their moons orbited the sun.

Astrologers continue to cast horoscopes, believing in the mysterious forces that they claim act upon a person at birth and determine their destiny and future. Astrology survives on the faith of people believing that their futures are preordained by cosmic forces over which they have no control. But astronomy has evolved into a true physical science, with verifiable observations and predictions based on the known laws of nature. Astronomers continue to study the physical laws of the universe, expanding their arena of observation out beyond the solar system to our galaxy and the universe.

The history of astronomy can be divided into three periods: The geocentric, the heliocentric, and the universal. The first had its beginnings in ancient civilizations and came to a close in the sixteenth century. The second extends from the seventeenth through the nineteenth centuries. The third began in our present century and continues today.

The Geocentric Period

Early astronomers believed that the Earth was at the center of the universe and assumed that the Sun, Moon, and stars revolved around the stationary Earth. The belief that the universe was geocentric or "Earth-centered" arose naturally from the apparent motion of objects in the sky. Simply stated, objects rose in the east and then set in the west. This pattern repeated itself day and night. The interest of these early astronomers, hardly scientific in our sense of the term, was mainly in practical matters, in the real and supposed relation of celestial events to events and people on the Earth. The early astronomers searched the skies for clues to good and evil omens, and learned to predict eclipses and to interpret the appearance of comets and other celestial messengers in order to improve their horoscopes.

However fanciful and unscientific these goals seem today, some remarkable discoveries were made. For example, pre-Columbian Aztecs developed calendars of great accuracy. The ancient Babylonians created amazingly precise star charts, while the classical Greeks determined the true dimensions of the Earth and carefully defined the ecliptic, the apparent path of the Sun through the stars. Continuous observations beginning 4,000 years ago by

Chinese astronomers led to timetables of the complete cycle of solar and lunar eclipses. As early as the second century B.C., early Greek observers discovered that the Earth wobbles as it spins on its axis—a phenomenon called precession.

The Heliocentric Period

The Polish astronomer Nicolaus Copernicus (1473–1543) is closely associated with the end of the geocentric period and the genesis of the heliocentric period in the sixteenth century. Astronomy as a science can be said to have begun in this period, divorcing itself from astrology. Copernicus reviewed the existing astronomical theories that led him to propose a new model for the Solar System. Most importantly, he placed the Sun, rather than the Earth, at the center of all things. He set forth his reasoning in his *De Revolutionibus* (*On the Revolutions of the Heavenly Spheres*), published in the year of his death.

Although not all of his astronomical ideas have proven to be correct, he is nonetheless credited with championing the accurate heliocentric, or Sun-centered, model of our Solar System. Copernicus demonstrated that, far from being the center of the universe, the Earth was merely the third in line of the six known planets revolving around a central Sun. And like the other planets, it traveled through its celestial motions in an ordinary way, governed by the known forces of nature. This revolution in astronomical thought is now called the first Copernican revolution and we shall see that it was followed by two others in the next centuries.

Soon new and even more radical ideas followed. Astronomers realized that the Sun was not the center of the universe, but one of billions of stars in the heavens. This is sometimes referred to as the second Copernican revolution. Some of these stars are larger, smaller, or even a different color than the star we call our Sun. In fact, it was agreed that there is no center to the universe. Astronomers soon believed that the universe is infinite in all directions and that nature's laws work in the same way throughout the universe. This is sometimes referred to as the third Copernican revolution.

In this period astronomy became increasingly scientific, motivated largely by the desire to know and understand the basic laws governing the motion of heavenly bodies. There was a desire to explain what the eye actually saw as opposed to unquestioning acceptance of earlier models of the universe that tried to fit the sky and its motions into perfect spheres and wheels. Progress from the sixteenth through the nineteenth centuries resulted from the effective combination of extended observations, improved instruments, and the work of a remarkable number of brilliant scientists.

Great quantities of fundamental and important data were painstakingly gathered by careful observers, chief among whom was the Danish astronomer Tycho Brahe (1546–1601). Without the aid of the telescope, Tycho gathered enough information about the orbit of the planet Mars to allow Kepler to form the laws of planetary motion. Tycho's observations were so accurate and detailed that he determined problems in Copernicus' theory of planetary motions were due to errors in observation—not the theory itself.

The introduction of the telescope in 1610 by Galileo Galilei (1564–1642) was a milestone in the science of astronomy. Another important invention was the spectroscope, developed by William Wallastron (1766–1828) and Joseph von Fraunhofer (1787–1826). These two instruments complement each other: The telescope permits us to see the stars more clearly and to collect more light from faint objects; the spectroscope analyzes the starlight, furnishing us with information about the temperature, composition, motions, and the age of individual stars.

Like every science, astronomy requires great minds that are able to apply insight, imagination, and intuition in order to advance. Johannes Kepler (1571–1630) and Sir Isaac Newton (1642–1727) were two such scientists during this period. Kepler, with excruciating precision, used Tycho Brahe's observations of Mars to discover the laws of planetary motion. Newton elegantly refined Kepler's laws and described the motion of the Moon with his own revolutionary formulation of the Universal Law of Gravitation. Simply stated, Newton mathematically demonstrated that the combination of the forward motion of the Moon and the downward pull of the Earth's gravity were what kept the Moon in orbit around the Earth. Newton's discovery also helped to explain the motions of the planets around the Sun.

The Universal Period

Early in this era, it became apparent that the galaxy of stars to which our Sun belongs is merely one of millions of galaxies—some larger or smaller than our own. Much of the astronomical research of the last half century has been devoted to studying these "island universes," in an effort to achieve a more accurate picture of the large-scale universe. To aid in this research, ever greater ground-based optical and radio telescopes have been built. Manned and unmanned space missions have also used intricate telescopes, including instruments that scan the heavens to receive wavelengths of energy (such as infrared or radio waves) not visible to the human eye.

The theoretical genius most closely associated with this period in astronomy (although he was primarily a physicist and mathematician) is Albert Einstein (1879–1955). Astrophysics, and cosmology in particular, depend more and more on his general theory of relativity developed in the early twentieth century. With the recent discoveries of gravitational lenses, and the almost certain existence of black holes, astronomers can no longer use classical Newtonian physics to describe correctly the properties of the universe. They must apply Einstein's relativistic physics to explain the results obtained with the new generation of instruments.

Astronomy is an exciting science, and like all sciences, continues to evolve and change as our knowledge of the universe grows. Astronomy uses the other sciences, such as physics (especially high-energy particle physics), chemistry, biology, and even higher mathematics to explain the wealth of fascinating information that is being collected by astronomical instruments. In fact, information from the two

Voyager missions to the outer planets (1977–1989), the 1987 supernova in the Large Magellanic Cloud, and the Infrared Astronomical Survey will provide astronomers with enough data to analyze well into the next century. Already newer and more sensitive instruments are probing the deep reaches of space, providing us with even greater knowledge of our universe and some new mysteries.

CHAPTER ONE

An Inventory of the Universe

KEY TERMS FOR THIS CHAPTER

astronomical unit (AU)
dark matter
ecliptic
elliptical
galaxy
light-year
nebula
nuclear fusion
parabolic/hyperbolic
plasmas
primordial
protostar
satellite

A Brief History of the Universe

The most reliable theory to date for the history of the universe is the one known as the big bang theory. According to this theory, all the matter and energy that is present in the universe today was once concentrated in a very small and enormously hot, dense ball. Approximately fifteen to twenty billion years ago, the ball exploded (the big bang!). The explosion created space and sent floods of energy in every direction. The energy immediately began to cool. After about 700,000 years, it had cooled enough to separate into a mixture of both energy and various types of hot, gaseous matter.

As time went on, concentrations of matter formed in the turbulent gases. Each concentration of matter began to contract under the influence of gravity while, at the same time, expanding away from other matter near it because of the initial explosion. Massive concentrations became galaxies and clusters of galaxies, while smaller clumps within the galaxies formed **protostars,** masses of gas and dust destined to ignite into stars. Many of these spinning protostars, while shrinking and flattening under the influence of their own gravity, gave rise to protoplanets and to protosatellites.

Protostars collapsed to become stars. The protoplanets and protosatellites, after a period of cooling, condensing, and contraction, became planets and satellites. To the best of our knowledge, the formation of our Sun from a protostar took place some 4.6 billion years ago. The planets and the satellites of the Solar System formed at the same time or shortly thereafter.

The Scale of the Universe

Astronomers have noticed a hierarchy of smaller systems to larger: The planets, asteroids, satellites, comets, and meteoroids revolve around a single star—the star we call the Sun. Together they make up the Solar System. The Sun and billions of other stars are members of the community of stars we refer to as our galaxy or the Milky Way galaxy. The universe contains many such stellar communities, with galaxies forming clusters and eventually superclusters.

Distances between stars are much greater than the distances between the planets. Distances between galaxies are still greater than distances between the stars. In attempting to understand such extraordinary distances it is essential to use some sort of scale. The scale that is commonly used represents the average Sun–Earth distance or 93 million miles. This measurement is called an **astronomical unit (AU).** One AU is equal to one foot (1 AU = 1 foot). On this scale, Mercury is four-tenths of a foot from the Sun. Venus is seven-tenths of a foot, and the Earth is one foot from the Sun. Pluto is then forty feet from the Sun, and a circle with a radius of forty feet would accommodate all the planets. This circle would be flat with all of the planets orbiting in approximately the same plane.

An Astronomical Yardstick

The average distance from the Earth to the Sun is 93 million miles. The nearest star, called Alpha Centauri, which is actually a triple-star system, lies some 25,000,000,000,000 (25 million million) miles away. The majority of other stars in our galaxy are even more remote.

The mile is not a useful unit of measurement when we deal with the immense distances between stars and galaxies. Instead, astronomers have adopted a larger "yardstick" called the **light-year.** A light-year (abbreviated "ly") is the distance a beam of light travels in one year. It is important to realize that it is a measure of distance and not time. Let us see how far that distance really is: The distance covered by a beam of light in one second is 186,000 miles. Hence

One light-year = 186,000 x 60 (seconds/minute) × 60 (minutes/hour) x 24 (hours/day) x 365 ¼ = 5,880,000,000,000 miles.

We can calculate that by accounting for all the seconds in a year, the beam of light will travel 5,880,000,000,000 or approximately 6 million million miles in one year.

Thus, the nearest star to the solar system, Alpha Centauri, is 4.35 light years away. For comparison, the diameter of our galaxy is about 100,000 light-years. Its maximum thickness at the center is 15,000 light-years. An average distance between galaxies would be approximately one million light-years.

The Sun is only a tiny fraction of a light-year from the Earth. Therefore, the distance to the Sun may be more conveniently expressed in terms of light-minutes. Since it takes a beam of sunlight eight minutes to reach the Earth, the Sun is eight light-minutes from the Earth.

Distances to heavenly bodies, when stated in terms of the speed of light, take on an added meaning. If a star such as Alpha Centauri is 4.35 light-years away, then that beam of light that reaches our eyes today began its journey 4.35 years ago. The most distant object visible to the unaided eye is the Andromeda Galaxy, 2 million light-years away. The light reaching Earth today left the galaxy when humanity was in its infancy, 2 million years ago. We cannot know what the Andromeda Galaxy looks like today for another 2 million years, until the light presently leaving it has arrived at the Earth. In other words, astronomers who observe such distant objects are actually looking at light that is very old.

An Inventory

What objects are easily seen in the sky? The unassisted eye is capable of detecting one nearby star, the Sun; one satellite, our Moon; five planets; shooting stars or meteors; several thousand stars; three neighboring galaxies; and an occasional comet. These celestial bodies represent the many objects in the universe in much the same way that homes, schools, hospitals, and parks make up a community.

To the best of our knowledge, the matter in the universe consists of galaxies and clusters of galaxies, billions of stars, clouds of dust and gas called nebulae, planets, asteroids or planetoids, satellites, comets, and a variety of smaller particles. Recent research has been also focused on the search for so-called **dark matter.** Dark matter is an undetermined form of matter that so far remains undetectable and may make up the bulk of the universe.

Galaxies and Clusters

Galaxies are huge, gravitationally bound collections of stars, dust, and gas. They are among the largest physical structures in the universe. The stars of our own galaxy, the Milky Way, have been estimated to number close to two hundred billion. This number can be written as 2 times 10^{11}, or 2 with 11 zeros after it.

A galaxy is classified in terms of its appearance as a spiral (as our own Milky Way Galaxy is believed to be), an elliptical, or an irregular galaxy. From the top, our galaxy would look circular in shape, with spiral arms resembling a pinwheel. A side view would look similar to a lens, or thick in the center and thin toward the edges. Again using the one-foot scale explained above, the diameter of the circle would be close to a million miles, while the maximum thickness would be only one-sixth of the diameter, or 170,000 miles. An elliptical galaxy would appear like a thick lens without the spiral arms; and, as the name implies, an irregular galaxy has no well-defined shape.

Galaxies contain hundreds of millions of stars, in addition to raw materials—mostly dust and gas—from which the stars are formed. However, the individual stars are so widely separated that if two galaxies were to collide, the galaxies would virtually pass through each other!

Ours is not the only galaxy in the universe. Many others have been photographed and studied in recent years, some strikingly similar to our own. Two examples of galaxies that resemble our own are M31, or the Andromeda Galaxy that is some 12 million trillion miles from us, and the spiral galaxy M51, or the Whirlpool Galaxy in the constellation Canes Venatici that is some 90 million trillion miles from us. Using our one foot equals one AU scale, the distances between galaxies are from ten to twenty million miles.

Galaxies are also found in groups of a few individual to several hundred members. These groups are called clusters and were once believed to be the largest physical structures in the universe. Some clusters contain only a few galaxies. Our galaxy, the spiral galaxies M31, M33, and the Large and Small Magellanic Clouds travel together and form a cluster called the Local Group. The latest information from orbiting telescopes indicates that clusters can form larger groupings called superclusters that contain many hundreds or thousands of individual galaxies. An example of such a large cluster is the giant group of galaxies in the direction of the constellation Hercules. Astronomers now have evidence that clusters of galaxies form large strings and bubbles that stretch across millions of light-years of space.

Stars

Stars are large spheres of gas that are intensely heated and shine through the process of **nuclear fusion.** This process rapidly changes hydrogen into helium, and thereby releases tremendous amounts of energy. At their surface, stars reach temperatures of thousands of degrees; in their interiors, temperatures are much higher, some reaching millions of degrees. At these temperatures, matter cannot exist as a solid or liquid, but only as a gas. Star gases are much denser than those found on the Earth and are called **plasmas.** Their extremely high densities are due to the enormous pressures that exist in the interiors of stars.

Stars move about in space, although their motions are not easily perceived. Usually no change in their relative position can be detected in a year or even a thousand years, except for a few very close stars such as Barnard's Star. The apparent lack of stellar movement is due to the vast distances separating us from the stars.

Distances to stars are much greater than distances to planets. The nearest star is 270,000 AU from our Sun or a distance of fifty miles on the AU scale! The majority of the stars are much further away. At such distances it takes many thousands of years for the stellar positions to undergo a noticeable change.

Distances between planets are stated in feet, while those between stars are stated in miles in our scale model of the universe. A mental picture might help to visualize this distinction: If the Sun and all the planets were put in a sphere with a radius of 40 feet, the closest star would be in a sphere 50 miles in radius. By our scale, other stars are at distances of thousands and hundreds of thousands of miles from the Sun!

Nebulae

Nebulae, from the Latin word **nebula** meaning "a cloud," are vast clouds of dust and gas found all over the universe, but they are primarily found in the star-forming regions of galaxies. The gases that form nebulae are extremely thin and have low temperatures. There are three major types of nebulae. A reflection nebula, such as the dusty patches around the young cluster of stars known as the Pleiades, does not shine by its own light but is visible because of the reflected light from neighboring stars. An emission nebula, such as the Great Nebula in the constellation Orion, is heated by a star or stars embedded in it and glows much like a fluorescent light. To the unaided eye, an emission nebula appears to be a fuzzy star. Its actual size and structure is only revealed with the aid of optical and radio telescopes. Dark nebulae neither reflect light or glow, but obscure the bright stars behind them, often creating interesting back-lit shapes such as the Horsehead Nebula in the constellation of Orion.

The Solar System

The Sun and the planets are the major components of the Solar System. Other members of this system are the host of smaller objects known as asteroids or minor planets; the planets' moons, known as **satellites,** that revolve around seven of the planets (not to be confused with artificial satellites); comets that appear from time to time; and a vast number of meteoroids.

The Sun. Although it may not seem so to us, the Sun is just an ordinary star, similar to the numerous stars we see in the sky. The Sun appears large to us, relatively speaking, because it is so near. All the other stars appear as small points of light in the sky because they are so far away (Figures 1.1 and 1.2). We are interested in our star, the Sun, because the Earth receives all life-sustaining energy from it.

Fig. 1.1. The sun is just an ordinary star. All the other stars look tiny, as they are so remote that we see them only as mere points of light.

Planets. There are nine planets revolving around the Sun: Mercury, Venus, Earth, Mars, Jupiter, Saturn, Uranus, Neptune, and Pluto, in that order from the Sun. The Earth's distance from the Sun varies because its orbit is elliptical, but on the average, it is 93 million miles. Mercury is only four-tenths the Earth's distance or .4 AU from the Sun. Pluto, the most distant planet, is forty times the Earth's distance or 40 AUs from the sun.

The planets that revolve around our Sun are either spherical rocky masses with slightly molten, solid cores, as in the case of the terrestrial planets of the inner solar system (the space in our Solar System from the Sun out to the orbit of Mars); or, they are huge envelopes of gas which surround liquid, metallic cores of hydrogen under extreme pressures, as in the case of the four-gas giant, or Jovian, planets of the outer Solar System (the space in our Solar System from the orbit of Jupiter to beyond

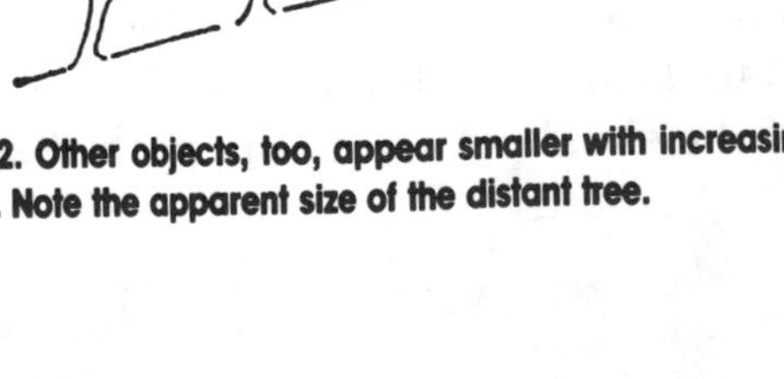

Fig. 1.2. Other objects, too, appear smaller with increasing distance. Note the apparent size of the distant tree.

the orbit of Pluto). The diameters of the planets range from less than 2,000 miles for Pluto to 86,000 miles for Jupiter. The planet most known to us is, of course, our own Earth. Unlike stars, all of the planets are relatively cool and shine only by reflected sunlight. The unaided eye can see several planets depending on the brightest planets' positions in the night sky. Uranus, Neptune, and Pluto are only visible with the use of a telescope.

At first glance, planets look much like the stars that glitter in the sky. But an observer can identify a planet by one or more of the following characteristics:

1. Planets shine with a steady light except when they are near the horizon. But light from stars seems to change rapidly in both color and brightness, creating a twinkling effect. This twinkling is caused by the various layers of Earth's thick atmosphere.
2. Planets wander in the heavens. A planet that appears in one constellation will move slowly into another constellation over the course of a few days or months. Stars, on the other hand, seem to keep the same positions relative to one another. The very word "planet" comes from the Greek word for "wanderer."
3. Planets appear as definite disks of light when observed through a telescope. The greater the magnification, the larger the diameter of the disk appears in the eyepiece. Stars will always appear as points of light even in the most powerful telescopes and never seem to have any measureable diameter.
4. Planets are only found in a narrow strip of the sky, fairly near the path that the Sun follows—called the **ecliptic.** Stars, of course, are found all over the sky.

Asteroids, Satellites, Comets, and Meteoroids

Asteroids are small, irregularly shaped solid bodies that also revolve around the Sun. They are also known as either planetoids or minor planets. The largest asteroid, Ceres, has a diameter of 480 miles. Three other asteroids—Pallas, Vesta, and Hygiea—have diameters greater than 200 miles. The majority of the asteroids have diameters of only two miles or less. Most asteroids are confined to the asteroid belt, located between the orbits of Mars and Jupiter.

The first minor planet was discovered on January 1, 1801, and thousands more have since been found. Like the larger planets, asteroids shine by reflected sunlight but are not very bright because of their small surface area. The brightest minor planet, Vesta, is visible to the naked eye when it comes close to the Earth.

Seven of the nine major planets have one or more satellites revolving around them. These satellites are more commonly known as moons. The Earth has only one satellite, our Moon, while both Jupiter and Saturn have satellites in the double digits. To date some 60 satellites have been discovered, largely as a result of the Voyager missions that began in 1977 and ended in 1989. The images and data from these spacecraft have shown astronomers that every satellite has had its own history of geologic evolution.

Comets are unique among celestial bodies in that they are visible only as they approach the Sun. They represent the original (or **primordial**) material left over from the formation of the Solar System. A typical comet consists of a luminous sphere, or head, connected to a long, tenuous tail. The tail develops and points away from the Sun, as the comet's gases react to the Sun's heat and solar wind.

To the naked eye a comet appears as motionless as the Moon. Actually, it moves at speeds of hundreds of miles a second, and its position can be seen to change each night when compared to the background stars. Its exact speed also can be determined by its relative position against the background stars. There are hundreds of comets with reliably known orbits and new comets are discovered every year. However, most comets are difficult to spot because they are too faint to be seen with the unaided eye.

The orbits of comets are either **elliptical** (closed ovals) or **parabolic** or **hyperbolic** (open ovals). The comets in elliptical orbits (orbits much like those of the planets) are regular return visitors to the inner Solar System, such as Halley's comet that appears every 76 years. Others that follow parabolic or hyperbolic orbits appear once and then head back into deep space, never to be seen again.

Meteoroids are actually tiny (about the size of the head of a pin) solid objects travelling through the solar system. Occasionally a group of meteoroids crosses the Earth's path and enters the atmosphere. Because they are moving very fast through the thick atmosphere, they literally burn up from friction. The streak of light that results from this fiery entry into the atmosphere is called a meteor or "shooting star," with the glow lasting several seconds to minutes. When many meteoroids enter the atmosphere during one or two nights, it is called a meteor shower.

Hundreds of tons of meteoric dust fall to the Earth's surface each year. On rare occasions large meteoroids manage to reach the Earth's surface before they are completely vaporized. These surviving fragments are called meteorites and range in size from small pebbles to giant boulders weighing several dozen tons. The first undeniable evidence that "space rocks" had landed on the Earth was on April 26, 1803, in France, when the startled population of L'Aigle was showered with thousands of meteorites. When the citizens picked up the space rocks, some of the pieces were still warm.

The visible matter in the universe is bound up in a number of objects that include a vast range of sizes—from the largest superclusters of galaxies to the smallest dust particles. All of this material came from a common origin and has been mostly influenced by the universal force of gravity. In the chapters that follow we will examine each of these objects and try to understand how they evolved.

CHAPTER TWO

The Earth and the Moon

KEY TERMS FOR THIS CHAPTER

albedo	*focus (pl., foci)*	*perigee*
apogee	*fronts*	*precession*
axis	*ionized*	*sidereal, synodic*
celestial sphere	*nodes*	*umbra*
ellipse	*nutation*	*zenith*
equinoxes	*penumbra*	*zodiac*

The Earth

In this chapter we will examine the Earth-Moon system, a system that has been studied very closely. The Earth is one of the smaller of the nine planets that revolve around the Sun. It ranks fifth in diameter, fifth in mass, and third in distance from the Sun. (For more basic data, see Appendix A.) It is very similar to several of the other planets in the solar system except that it is the only place in the universe that is known to sustain the existence of life.

As a base for astronomical observations, the Earth is far from perfect because it is not stationary. All observations must be corrected for the Earth's movements. Nor is the movement of the Earth simple; it is a highly complex combination of at least six different and simultaneous motions:

1. Once a day, it rotates about its axis, the imaginary line through its center.
2. The axis (along with the rest of the Earth) revolves around the Sun (once a year).
3. The axis undergoes precession, or wobbling.
4. The axis undergoes nutation, or nodding.
5. The Sun, with the Earth and the other planets, is speeding through the local

cluster of stars toward the bright star Vega at 12 miles per second.

6. The local cluster of stars takes part in the rotatation about the center of our galaxy at an average speed of 200 miles per second.

Our senses do not make us aware of these motions, much as passengers on a smoothly running train are hardly aware of the train's speed. Only when a passenger looks outside the window and notices the moving scenery is the true speed of the train detected. This is much the same situation with an observer on Earth. We must use other celestial bodies as references to appreciate the true motions of the Earth.

The Earth has one natural satellite, the Moon, which continuously revolves around it, while the Moon makes its own journey around the Sun. Although the Moon's volume is only 1/50 and its mass 1/81 of the Earth's, it affects the Earth in a rather striking way. The periodic rise and fall of the oceans, or tides, are a direct result of the gravitational pull of the Moon.

The extreme closeness of the Moon has made it easy to study since the early days of astronomy. The average distance between the Earth and the Moon is merely 60 times the Earth's radius, or 238,000 miles.

Some data about the Moon has been gathered by telescope, but much of what is known has been learned from naked-eye observations. For example, it is well known that the Moon goes through a complete set of phases every month, from new to crescent, to quarter, to full. In addition, the Moon moves in almost the same apparent path in the sky, known as the ecliptic, as the Sun and the other planets. It is also known that the Moon rises each day an average of 51 minutes later than on the previous day, and that part of the Moon's surface always faces the Earth while part is always hidden.

Rotation of the Earth

The Earth spins counterclockwise about its axis. It completes one revolution in a sidereal day, which is 23 hours, 56 minutes, and 4 seconds long. (The sidereal day is shorter than the "clock" day, which is known as the "mean solar day.") Experimental proof of the Earth's rotation was first demonstrated by the French physicist J.B.L. Foucault in 1851. By setting into motion a free-swinging pendulum suspended by a long wire from the ceiling of the Pantheon in Paris, he gave graphic proof that the Earth turns on its axis. Foucault traced a line on the ground marking the direction of the pendulum's swinging bob. He observed that one hour later this mark had turned 15° in a counterclockwise direction relative to the plane in which the pendulum was swinging. In one sidereal day the line had rotated 360°.

There is other evidence of the Earth's rotation. When artificial satellites were launched into Earth orbit beginning in the late 1950s, these satellites were able to pass over different parts of the Earth because the Earth was constantly turning under them. This has also been observed in all of the hundreds of satellites that have been launched since that time. Other satellites have been placed in stationary, or geosynchronous, orbits where the speed of the satellite's revolution around the Earth exactly matches the rotation of the Earth.

The net result is that the satellite remains over one point of the Earth's surface at all times.

We are not aware in our daily lives of the true motion of the Earth. What is observed is an apparent motion; that is, it seems that the celestial sphere is turning above our heads. Our eyes tell us that the Sun and the stars rise in the east, travel in a long arc above us, and then set in the west. This relationship between true and apparent motion has its counterpart in a moving train: Looking from a window of a north-bound train, one sees the apparent south-bound motion of the landscape.

Several effects are directly due to the spinning of the Earth:

1. Every place on the Earth alternately faces toward the Sun (day) and away from the Sun (night).
2. The Earth's axis maintains its inclination to the plane of its orbit (66½°), always pointing to the North Star. In this respect, the rotating Earth resembles a child's spinning toy top. Like a top, the Earth's axis precesses or wobbles.
3. A centrifugal force—stronger at the equator and zero at the poles—acts upon every object on Earth. This accounts for the fact that objects weigh more at the poles and less at the equator. (The difference in weight is tiny and is only primarily of scientific interest.)
4. The flattening of the Earth at the poles was probably caused by its spinning at a time when its surface was still liquid or at least plastic and indicates that the outer core of the Earth is still molten.

The Shape of the Earth

The Earth is nearly spherical in shape; the sphere is only slightly affected by its mountains and valleys. In fact, if it were scaled down to the size of a billiard ball, the Earth would be perfectly smooth.

Nonetheless, the Earth is slightly flattened at the poles: the polar diameter is 27 miles less than the diameter at the equator. Earth scientists call the shape of our planet an oblate spheroid. Data obtained from artificial satellites show that the Earth is actually pear-shaped with the South Pole indented and the North Pole slightly extended.

The oblateness of the Earth is also responsible for the variation of weight in latitude. Because an object at the pole is closer to the center of the Earth, it weighs more than at the equator. The change in weight from equator to pole is about ½ of 1 percent.

Revolution of the Earth

The Earth revolves around the Sun in a counterclockwise direction as seen from above the Sun's north pole. The orbit of the Earth is shaped like an **ellipse,** or flattened circle, with the Sun as one of its **foci**—the two central points of every ellipse.

Again, we do not see the true motion of the Earth's revolution, but rather the apparent motion of the Sun, which seems, in the course of a year, to make one revolution about the Earth. This apparent orbit of the Sun is called the ecliptic; a belt 8° wide on each side of the ecliptic is called the

zodiac. There are 12 prominent constellations within the belt: Aries, Taurus, Gemini, Cancer, Leo, Virgo, Libra, Scorpio, Sagittarius, Capricorn, Aquarius, and Pisces.

The Sun, in its annual path along the ecliptic, seems to pass through each of these zodiacal constellations, remaining in each for approximately one month. This apparent motion of the Sun through the 12 zodiacal constellations is used to define our sidereal year; that is, the time it takes for the Sun to complete one full circuit through the background stars. The sidereal year is 20 minutes longer than the tropical year, which is the year we track with our calenders.

Inclination of the Equator to the Ecliptic

The ecliptic, or the path the Sun follows, is inclined to the celestial equator at an angle of 23½°. The celestial equator is the outward projection of the Earth's equator onto the **celestial sphere,** or the imaginary bowl of the sky containing all the stars and centered on the Earth. (See Chapter 14 for more detail.) The ecliptic crosses the Earth's equator at two points, known as the **equinoxes:** The Spring (or Vernal) Equinox occurs in March, the Fall (or Autumnal) Equinox in September (Figure 2.1).

The inclination between the paths of the celestial equator and the plane of the ecliptic is the main cause of the seasons on Earth. When the Sun is north of the equatorial plane, it is warm in the northern hemisphere and cold in the southern hemisphere for two major reasons:

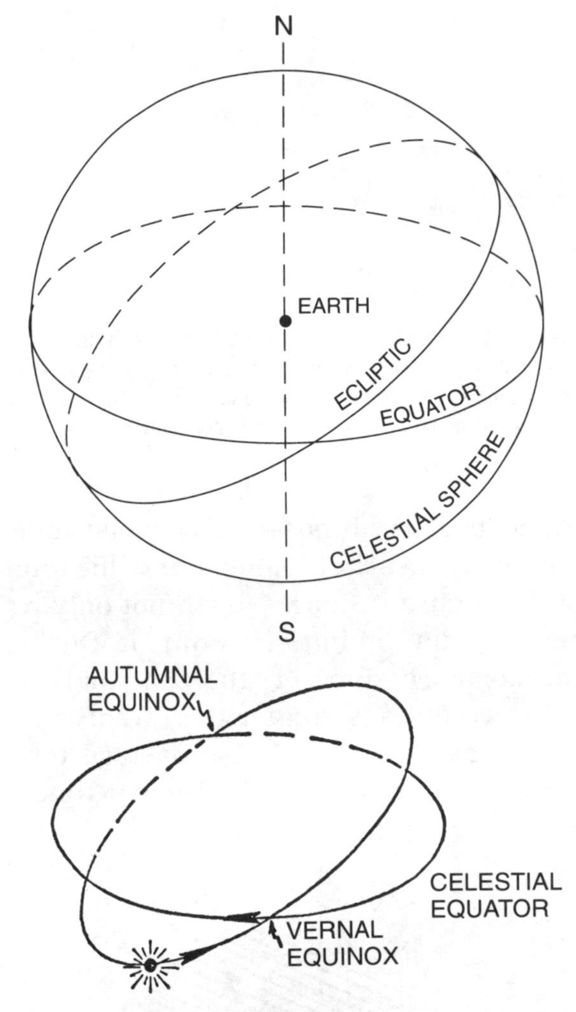

Fig. 2.1. The apparent path traced out by the Sun on the celestial sphere is known as the ecliptic. It is inclined 23°27′ with the Earth's (or celestial) equator. The points of intersection of the two orbits are called equinoxes.

1. The Sun's rays are more concentrated in the northern hemisphere, where the rays supply the heat for a relatively small area; in the southern hemisphere, the rays are spread over a much larger area. Consequently, the southern hemisphere receives less heat per unit area and is colder (Figure 2.2).

2. The Sun is above the horizon longer in the northern hemisphere. For example,

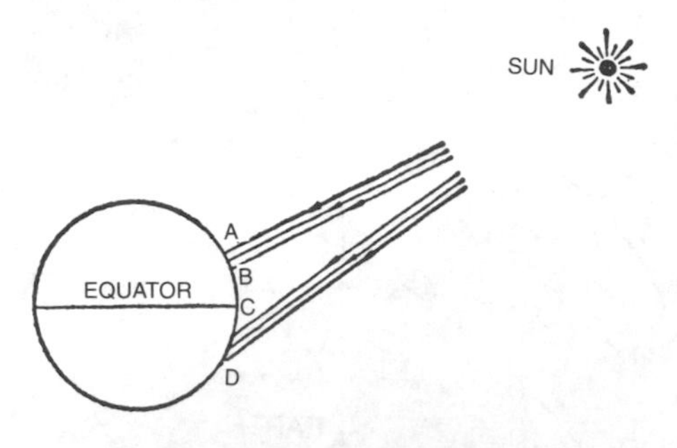

Fig. 2.2. The rays in the northern hemisphere are concentrated in the small arc AB. The same group of rays falling on the southern hemisphere are spread out over the large area CD. This is the primary reason for the warm season in the northern hemisphere.

in the month of June, at 40° north latitude, daylight is nearly 15 hours out of the total of 24 (Figure 2.3). The Earth not only receives radiation but also emits it. During the northern summer, the heat that the Earth receives is greater than what is radiated back into space, so average temperatures continue to rise for that time.

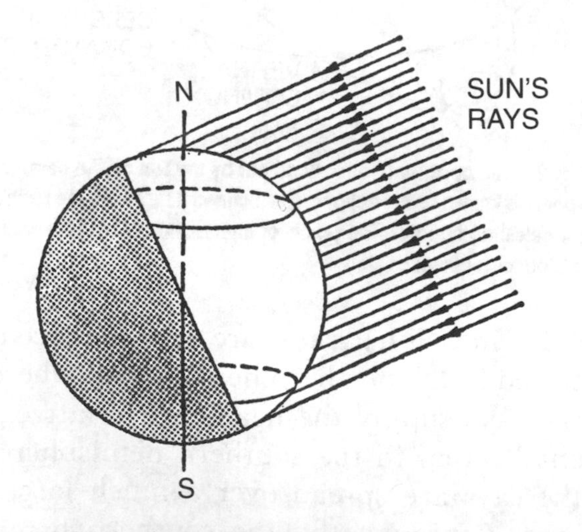

Fig. 2.3. Daytime is more than 12 hours in the northern hemisphere and less than 12 hours in the southern. This is a contributing factor to the warm season, while the Sun is above the celestial equator. On June 21 the Sun is farther from the equator. The hottest weather in middle latitudes occurs later: average temperatures in the United States, say, do not reach their height until late July or early August. This lag results from the balance of radiation.

The Earth, moving in its true elliptical orbit, is closest to the Sun during the southern summer. Due to this difference in distance, the southern hemisphere receives about 6 percent more solar energy than the northern hemisphere. This additional amount of heat for the southern hemisphere will not continue forever. In about 10,900 years, the northern hemisphere will receive that extra percentage of heat. Then, about 12,900 years later, it will again be the southern hemisphere's turn, and so on. This alternation is continuous and results from **precession** of the equinoxes, as the Earth's axis wobbles back and forth, completing a cycle in 25,800 years.

Precession of the Equinoxes

To understand the cause of the precession of the equinoxes, it is best to use the true and not the apparent motion of the Earth and the Sun. The Earth, in its elliptical orbit around the Sun, tilts its axis by 23½° from a line perpendicular to that orbit. As it revolves around the Sun and spins on its axis, the axis traces a small circle in the sky (Figure 2.4). At present the axis

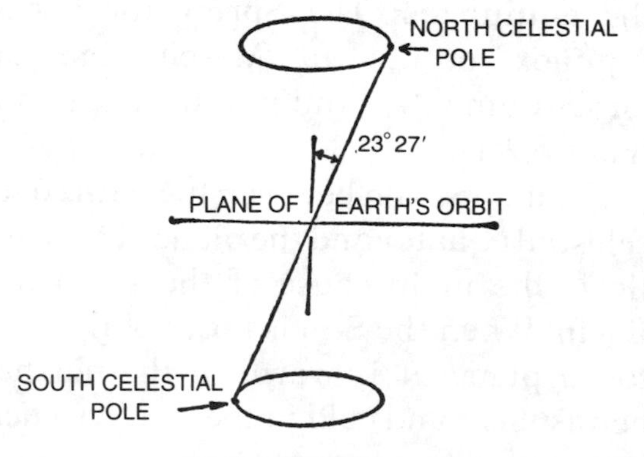

Fig. 2.4. The celestial poles trace out two circles each of 23°27′ radius.

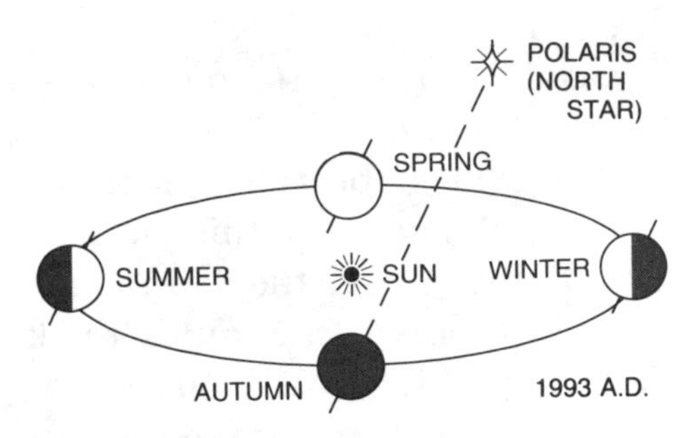

Fig. 2.5. The axis of the Earth at present points toward the North Star. On this picture the axis is "up to the right."

pierces the sky within 1° of Polaris, the star at the end of the handle in the constellation of the Little Dipper. Because Polaris is so close to the North Celestial Pole, it is also known as the North Star (Figure 2.5). The current position of the North Celestial Pole will change about the year A.D. 2100 when it will be directed to a point on the celestial sphere only ½° from Polaris. That is the closest the pole can be to Polaris.

About the year A.D. 14,000 the pole will be directed to a point 5° from the star Vega. At that time, the axis of the Earth will point to the left, still making an angle of 23½° with the perpendicular (Figure 2.6). Following that, the pole will continue to trace the small circle of 23½° radius in the sky, completing one revolution each 25,800 years.

The motion of the Earth's axis is caused by the pull of gravity, in the same way that gravity causes the leaning axis of a top to precess, describing the surface of a cone

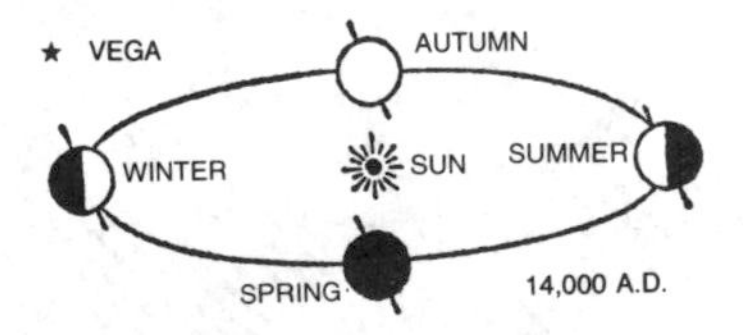

Fig. 2.6. In the years about 14,000 A.D. the Earth's axis will point "up to the left." Compare this to the present "up to the right" inclination. Note that the magnitude of the angle 23°27′ did not change.

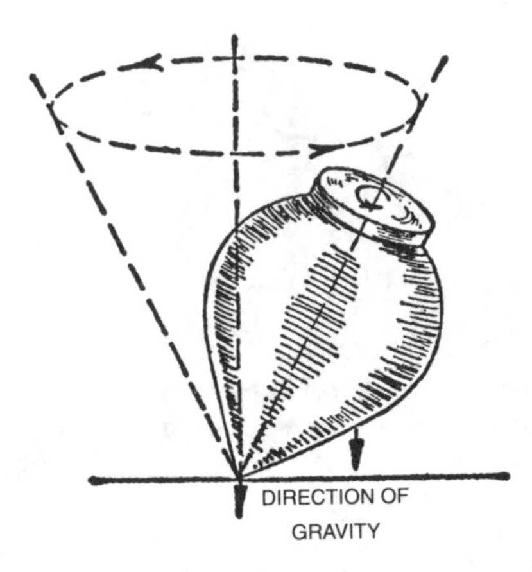

Fig. 2.7. In the case of a spinning top it is the force of gravity that causes the axis to precess, that is, to trace out a cone.

(Figure 2.7). The gravitational force that causes the Earth's axis to precess is exerted by the Sun and the Moon on the Earth's slight equatorial bulge.

The effect of this force is to change the direction of the axis, not its inclination. The axis remains tilted at 23½° while it traces the shape of a cone once every 25,800 years. The equinoxes move with the rotation of the axis, also making one revolution in 25,800 years.

Nutation

The curve traced by the Earth's axis is not a smooth circle. The true motion of the axis is a combination of precession and a nodding motion, or **nutation** (from the Latin word for "nodding"). The curve has small waves, due to the nodding of the Earth's axis about the average position of 23½°. The period of one such complete wave is 19 years; the nod at its maximum is 9 arcseconds (Figure 2.8). The gravitational pull of the Moon is the primary cause of nutation.

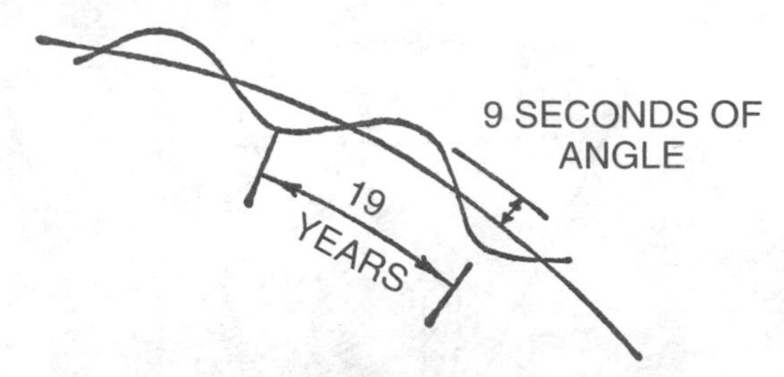

Fig. 2.8. The curve traced out by the North Celestial Pole is not a smooth circle. The waves in that curve are called nutations. These waves have a wavelength of 19 years and an amplitude of 9 arc-seconds.

The Interior of the Earth

With the new wealth of information obtained by the recent Voyager missions to the planets and satellites of our Solar System, astronomers have been looking to Earth's geology to help them explain the origins and dynamics of the Sun's other family members. The crust of the Earth is more than 30 miles thick under the continents, but scarcely more than 3 miles thick beneath the oceans. Direct observations of the interior of the Earth are available for only the top several miles.

Our knowledge of the layers beneath this crust is derived from the analysis of earthquake waves. Earthquakes are caused by the slipping of one part of the Earth's crust relative to the neighboring part of the crust. A crack in the crust is called a fault.

The arrival times of the various earthquake waves are carefully noted at geological observatories all over the Earth. Because waves travel at different speeds through materials of varying densities, the study of earthquake waves provides data to estimate the distribution of materials deep inside the Earth.

The results of this research indicate that the interior of the Earth can be divided into four parts: crust, mantle, outer core, and inner core (Figure 2.9).

The crust, the layer best known to us, has, as indicated above, a thickness of more than 30 miles under the continents, but is scarcely more than 3 miles thick under the oceans. Chemical analyses indicates that the crust is made up of about 47 percent oxygen, 28 percent silicon, 8 percent aluminum, 5 percent iron, and smaller percentages of a large number of the other elements. More recently scientists have found that the crust is made up of huge plates, or sections of the crust. This theory is often referred to as the plate tectonic theory. Connected to plate tectonics is the theory of continental drift, which states that the continents were not always in the same configuration as they are today. By looking at the contours of South America and Africa, one can see how these two widely separated continents were

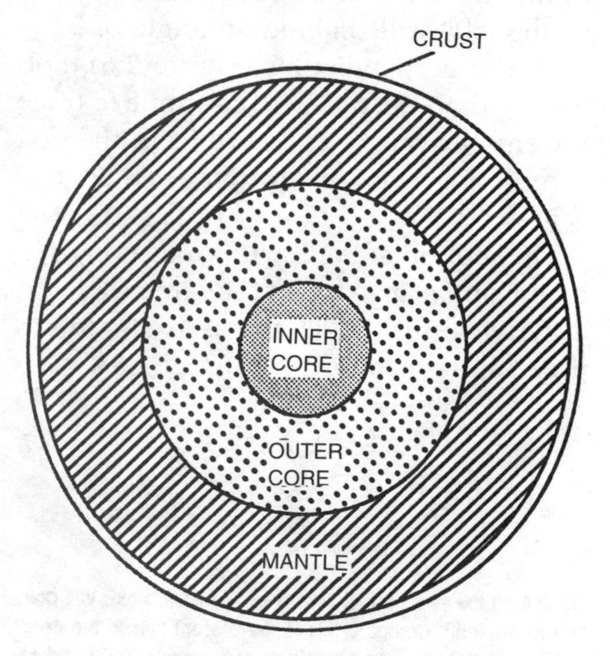

Fig. 2.9. The four layers of the Earth: crust, mantle, outer core, and inner core.

once part of a larger landmass, subsequently splitting and drifting apart.

Supporting the crust is a layer of dense rock 2,200 miles thick known as the mantle. Chemically, it is primarily made up of silicates that are rich in magnesium and iron. The Moho discontinuity, named for its discoverer, Yugoslavian seismologist Andrija Mohorovicic in 1909, is the uneven boundary between the crust and the mantle. The next layer is the outer core. It is more than 1,200 miles deep and most likely consists of nickel-iron. Studies of seismic waves traveling through this region indicate that the outer core is in liquid form. The inner core is the last layer and has a radius of about 800 miles. Like the outer core, the inner core consists of nickel-iron. However, the pressure at this region is so great that it is most likely solid.

The average density of the Earth is 5.5 grams per cubic centimeter, but the density of the material near the surface is 2.7 grams per cubic centimeter, or less than half the Earth's average density. (For comparison, the density of pure water is exactly 1 gram per cubic centimeter.) The assumption of nickel-iron in the core accounts for the high value of the Earth's average density. The rotation of this metal may also account for the Earth's magnetic field.

The Magnetism of the Earth

The Earth's magnetic field is similar to the field near a bar magnet. The axis of the Earth's magnetic field is inclined 12° to the Earth's rotational axis. The North Magnetic Pole is near Hudson Bay, while the South Magnetic Pole lies in Victoria Land in Antarctica. The locations of the magnetic poles migrate from time to time because the Earth's core is rotating at a slightly different rate than its outer layers.

Scientists have traditionally believed that the Earth's magnetic field is caused by the circulation of liquid metal in the outer core of the rotating Earth. Current research has found that both the Sun and the Moon have magnetic fields. Yet the Sun contains no iron, and the Moon does not have a liquid core. Discovering the sources of the magnetic fields of the Sun and the Moon may help scientists identify the source of the Earth's magnetic field.

We can see the evidence of the Earth's magnetism with a magnetic compass. The compass needle orients itself parallel to the local magnetic line. Artificial satellites and space probes have shown that the magnetic field around the Earth is much more complicated than first thought. The complication is due to the interaction between the solar wind (a stream of ionized particles emitted by the Sun) and the Earth's magnetic field. The result of this interaction is a field known as the magnetosphere, which extends about 50,000 miles in the direction toward the Sun. In addition, data obtained by the Explorer 10 satellite in 1961 revealed a 150,000-mile-long magnetosphere tail pointing away from the Sun.

The Van Allen Radiation Belts

The Earth is surrounded by two belts of high-energy charged particles, mostly high-energy protons and electrons that are trapped by Earth's magnetic field. These belts, called the Van Allen radiation belts, were named after the American physicist James A. Van Allen, who discovered their existence in 1958 from Explorer 1 satellite data. The belts were further verified by Pioneer 3 in 1958.

The Van Allen belts are doughnut-shaped. The innermost belt is at a distance of about 2,000 miles above the Earth's surface, while the outer belt is about 10,000 miles above the surface. The trapped high-energy protons and electrons spiral along the Earth's magnetic lines of force, then bounce from one hemisphere to the other during the several days of their capture.

During the early days of manned space travel, there was considerable concern for the safety of astronauts travelling through these belts. However, repeated passages through the belts by the Apollo, Skylab, and Space Shuttle missions have indicated no threat to human life.

The Earth's Atmosphere

Surrounding the Earth's surface is an envelope of air. It protects all life from excess ultraviolet radiation from the Sun, aids in the equalizing of temperature extremes, and allows for respiration by all living organisms.

Air is a mixture of gases. The composition, by volume, is 78 percent nitrogen, 21 percent oxygen, less than 1 percent argon, and trace amounts of carbon dioxide, water vapor, and other gases. These percentages are different in the upper air, with hydrogen and helium playing an important part in the composition of the air at heights above about 30 or 40 miles.

The Earth's atmosphere has no sharply defined upper limit. A study of meteors indicates the presence of air at levels up to 100 miles, while the study of aurorae—bright, colored bands of light high in the atmosphere—indicates some atmosphere at least 400 miles above sea level.

The atmosphere can be divided into five layers or shells, named here in increasing altitude for discussion's sake: troposphere, stratosphere, mesosphere, thermosphere, and exosphere.

1. *Troposphere* One of the functions of the troposphere, or our weather layer, is to adjust the temperatures at the Earth's surface. Solar energy is usually excessive at lower latitudes, and rather sparse at higher latitudes. The exchange of air between latitudes moves part of the excess heat to the cooler parts, and vice versa, with the aid of large masses of air. The boundaries between these large air masses are known as **fronts.** Most of the so-called bad weather takes place at the fronts, in the form of clouds, fog, and all types of precipitation.

The height of the troposphere varies with latitude, decreasing from 10 miles above sea level at the equator to 5 miles or less at the poles. The temperatures in the troposphere range from an average of 56° F at sea level to a value of −60° F at the upper limit. The tropopause is the dividing boundary between the troposphere and the next layer.

2. *Stratosphere* This layer extends about 40 miles above the troposphere. The temperature here stays, on average, at a constant 60° F for the first 10 miles up, then decreases to 32° F in the next 10 miles, and finally plummets to −160° F in the top 20 miles of the layer.

The stratosphere contains about one-fifth of the mass of the entire atmosphere.

The air currents in the stratosphere are primarily horizontal, or parallel to the Earth's surface. Chemically, the stratosphere is similar to the lower reaches of the atmosphere with two exceptions: First, there is less water vapor and second, there is more ozone.

Ozone (O_3) forms in the first 10 miles of the stratosphere through the action of the Sun's ultraviolet radiation on oxygen (O_2) molecules. It is this ozone layer that shields all life from the harmful effects of ultraviolet rays. Unfortunately, recent studies have shown an alarming depletion in this ozone layer above the North and South poles, most likely caused by harmful chemicals released into the upper atmosphere by our technological society.

3. *Mesosphere* This layer lies above the stratosphere and extends to an altitude of about 50 miles. In this layer the temperature decreases as the altitude increases. In fact, the mesosphere is the coldest layer of the Earth's atmosphere, with temperatures approaching −100° C (−212° F). The upper boundary of the mesosphere, the mesopause, is marked by an increase in temperature.

4. *Thermosphere* In the thermosphere, temperatures increase steadily with altitude. Nitrogen and oxygen atoms absorb solar energy, explaining temperatures as high as 2,000° C (3,600° F).

The lower region of the thermosphere, at an altitude of 50 to 330 miles, is often called the ionosphere. It contains a considerable number of **ionized** atoms, that is, atoms that have had one or more electrons added or stripped from them, as a result of being bombarded by ultraviolet and X-ray radiation from the Sun. There are four reasonably separate layers in the ionosphere, universally known as the D, E, F_1, and F_2 layers. All four layers are of great importance in long-range radio broadcasting. Multiple reflections between these layers and the Earth make it possible to send radio signals around the world. When there are especially large and powerful eruptions of solar flares or sunspots, these layers in the ionosphere are disrupted and radio transmissions are often interrupted.

5. *Exosphere* This is the last layer and continues into outer space. Here, atoms and some lighter molecules escape from the atmosphere entirely, but ionized particles are prevented from escaping by the Earth's magnetic field.

There is growing concern today about the health of the Earth's atmosphere. Of special interest is the gradual warming of the Earth due to the so-called greenhouse effect. Similar to what happens in a glass greenhouse, carbon dioxide and water vapor molecules let in the warming radiation of the Sun while preventing the heat from escaping. Industrial pollution adds carbon dioxide, putting a lid on our atmosphere. If precautions are not taken, such as the reduction of CO_2 emissions, Earth may one day experience conditions similar to the greenhouse effect on Venus. (See Chapter 4 for more on Venus.)

The Atmosphere in Astronomy

The Earth's atmosphere affects incoming radiation in several ways: reflection, absorption, diffusion, and refraction.

1. *Reflection* The phenomenon of twilight is a direct result of reflection by very small particles of dust and smoke in the atmosphere. These particles reflect the

rays from the rising or setting Sun back to the Earth, thus providing additional daylight. Astronomical twilight lasts until the Sun's center is 18° below the horizon.

2. *Absorption* The Earth's atmosphere is a selective absorber. It absorbs nearly 100 percent of some wavelengths of light but only partially absorbs others. Short ultraviolet rays are totally absorbed in the atmosphere, while visible light waves are only partially absorbed. Therefore, starlight reaching the eye is materially different from the light leaving the star. Selective absorption complicates the work of an astronomer. Because the study of stellar spectra tells astronomers virtually everything about stars, this difference must be considered (see chapter 9).

3. *Diffusion* This effect results from the scattering of light by individual molecules of air, with the amount depending on the color of the light. Blue light is scattered or diffused much more than red light. This selective scattering accounts for both the blueness of the sky and the red and orange of sunsets. Blueness results from upper-air molecules scattering blue in all directions. The Sun appears red when it is low in the sky because more blue light rays are scattered by the atmosphere. Thus more of the longer-wavelength red light rays reach the Earth's surface, giving the Sun its red color.

4. *Refraction* Light passing from interstellar space through the atmosphere is refracted or bent. The amount of refraction increases as the light approaches the denser layer closer to Earth. As a result, all celestial bodies appear higher in the sky than they really are. The amount of refraction is greatest near the horizon and decreases rapidly as it approaches the **zenith,** or the point in the sky directly overhead. Refraction permits the stars and the Sun to be seen shortly before they rise and for a short while after they set. Refraction also produces the twinkling of stars. The density of the air at various levels changes rapidly due to the prevailing winds at these levels. Starlight through these levels is refracted in amounts that vary from second to second. This accounts for the rapid jumping or twinkling of the stars.

The Moon

The Moon in Its Orbit

The Moon revolves counterclockwise around the Earth in an elliptical orbit, with the Earth as one of its foci. The point on the ellipse nearest the Earth is called the **perigee.** The distance between the Moon and the Earth at perigee is 221,463 miles. The point on the ellipse farthest from the Earth is called the **apogee,** 252,710 miles.

The plane of the Moon's orbit lies very close to the plane of the ecliptic, the apparent path of the Sun on the celestial sphere. It is tilted about 5° in relation to the ecliptic. The two points where the Moon's path crosses the plane of the ecliptic are known as **nodes.** They are not constant in space but move along clockwise along the ecliptic, completing one revolution in 19 years (Figures 2.10 and 2.11). The Moon takes around 27½ days to complete its orbit around the Earth.

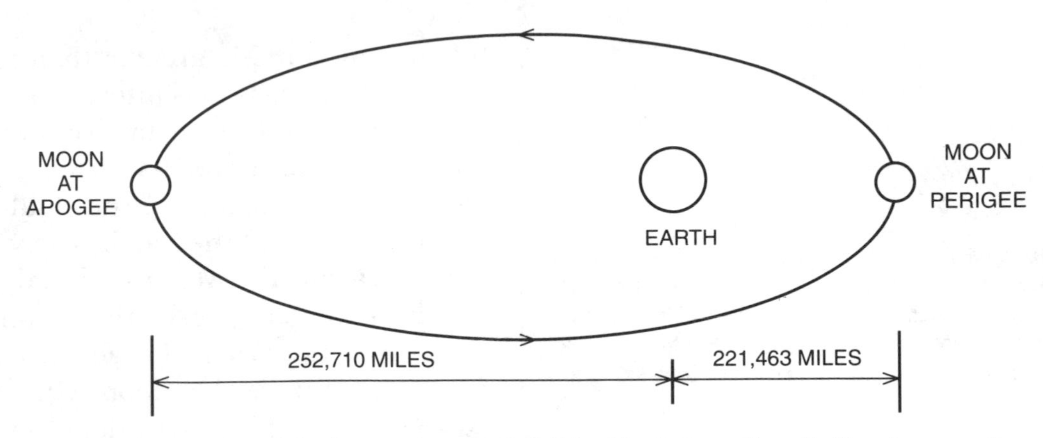

Fig. 2.10. The Moon in its orbit. An observer above the North Pole of the Earth would see the Moon trace an ellipse in a counterclockwise direction.

The Tides

The oceans of the Earth rise and fall in relation to coastal areas at more or less regular intervals, called tides. On an average, the period between two successive high tides is 12 hours and 25½ minutes—exactly one half the time it takes the Moon to complete its circuit around the Earth; or, one half of 24 hours 51 minutes. This is not a coincidence, as the ocean tides are caused primarily by the Moon's gravitational pull.

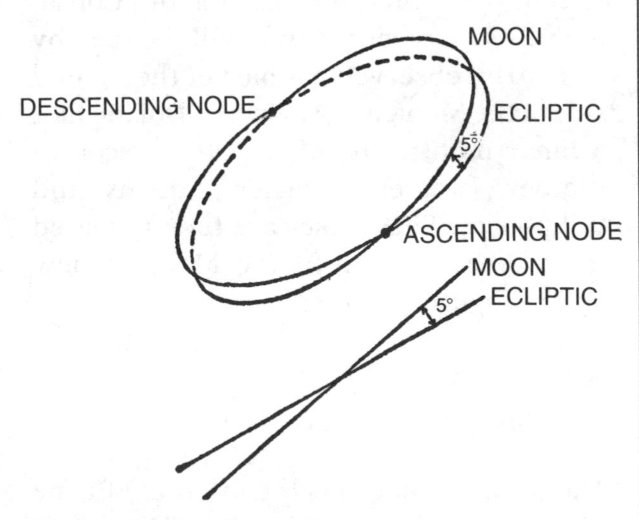

Fig. 2.11. The orbit of the Moon is inclined by 5° to the apparent orbit of the Sun (the ecliptic). This can be seen both in perspective (top) and in the side view (bottom).

Consider the center of the solid Earth, and a body of water facing the Moon, as well as a body of water on the opposite side of the Earth. The gravitational force of the Moon exerts a pull on all three. The intensity of the force, however, is largest for the water nearest the Moon. As a result, the ocean on the side of the Earth facing the Moon bulges slightly, causing a high tide within the area of the bulge. This is called the direct tide.

At the same time, another slightly smaller tidal bulge is created on the opposite side of the Earth. This is called the opposite tide. This bulge takes place because the ocean waters on the side opposite the Moon are not as affected by the gravitational pull of the Moon as is the solid Earth. Thus the ocean floor is pulled toward the center of the Earth and away from the water. The net effect is a tidal bulge.

Low tides are created halfway between the time of the two high tides. Low tides

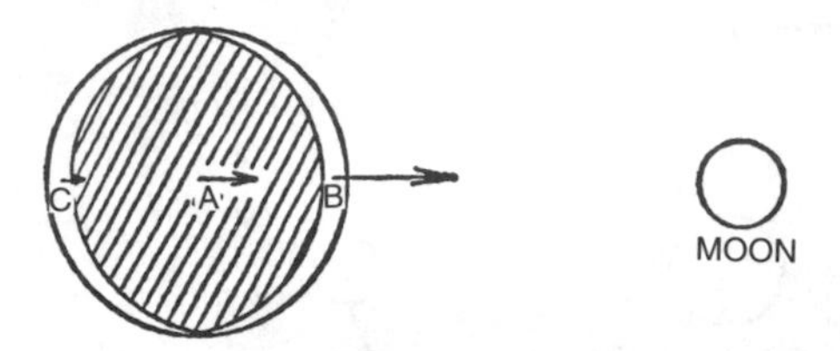

Fig. 2.12. The formation of tides. The force at A on the solid earth is larger than at C and thus the Earth pulls away (toward the right) from C, causing the high tide there. The force at B is larger than at A and thus the body of water at B pulls away from the solid earth causing the high tide on the side of the Earth facing the Moon.

occur because ocean water flows toward the areas of high tide (Figures 2.12 and 2.13).

The effect of the Sun on tides is secondary to that of the Moon—only about 7 percent—because of its much greater distance from the Earth. When the tide-raising forces of the Moon and the Sun do act together, however, the resulting tides are at maximum. This phenomenon occurs at new Moon when the Moon and the Sun are on the same side of the Earth. These tides are called Spring Tides. The other extreme is reached when the Sun is at 90° to the Moon. The tides are then at a minimum and are called Neap Tides.

The Moon's closeness to the Earth also has an influence on the strength of the tides. When the Moon is at perigee, the tide-raising force is greater than normal by some 20 percent. It is interesting to note that the Moon was once much closer to the Earth, but the Moon has been steadily moving away because of the tidal forces between itself and the Earth. These forces are increasing the Moon's orbital speed, which in turn, increases the Moon's distance from the Earth. In fact, it is likely that in the future the Moon will entirely escape the Earth's gravitational pull.

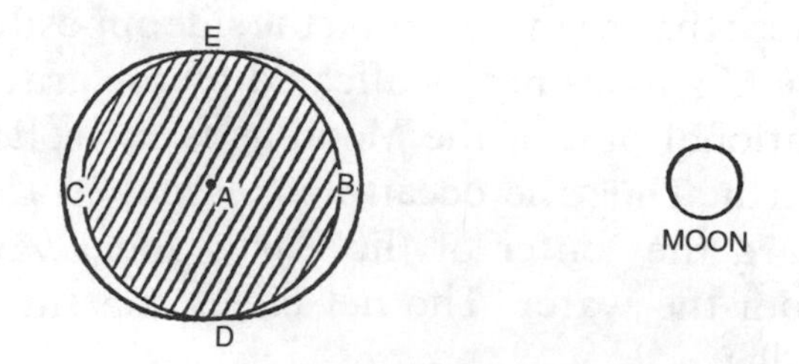

Fig. 2.13. The water at B is "escaping" from the earth causing high tide there. The water at C is being left behind by the "escaping" earth causing high tide there. The bodies of water at E and D will be at low tide at that time.

Phases of the Moon

The Moon seems to change in shape from a narrow crescent to a full face and again to a narrow crescent. This is because the Moon revolves around the Earth and receives its illumination from the Sun. At new Moon, the lighted side of our satellite is turned away from the Earth so the side facing the Earth is dark. Several days later, the Moon has moved from the position between the Sun and the Earth. Consequently, a crescent of light will be seen by terrestrial observers, as part of the lighted side of the Moon is now visible. During half a lunar month, the Moon can be seen in the new, crescent, quarter, gibbous, and full phases. The phases are then repeated in reverse order until the Moon is new again (Figure 2.14).

Two Kinds of Months

The **synodic** month is the interval of time between two consecutive new Moons, or two consecutive full Moons. It lasts for 29.53 days—or more exactly, 29 days, 12 hours, 44 minutes, and 2.78 seconds. As-

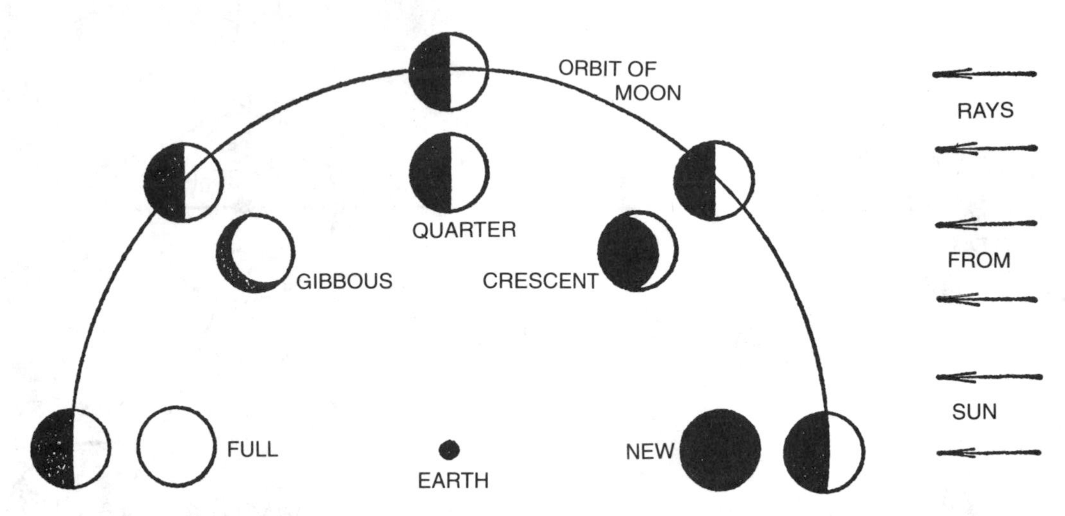

Fig. 2.14. The Moon in its various phases. The five circles along the orbit of the Moon indicate the lighted hemisphere on the Moon. The inner five circles indicate the corresponding illumination of the Moon as seen from the Earth.

tronomers base this definition on an alignment of the Sun, Earth, and Moon in a straight line, or rather, the interval between two such successive alignments (Figure 2.15). The synodic month is also known as the lunar month.

The sidereal month is the period of time elapsed between two consecutive times that the Earth and the Moon are in line with the same fixed star. The sidereal month, or star month, is shorter than the synodic month. Its period is 27.32 days, or 27 days, 7 hours, 43 minutes, and 11.47 seconds. The synodic month is longer be-

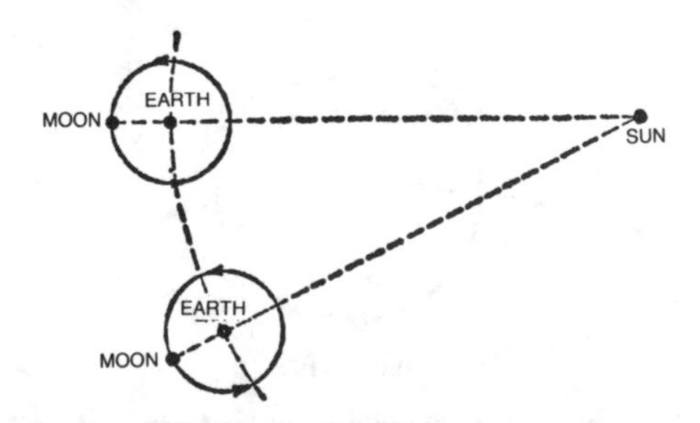

Fig. 2.15. The synodic month. It is the period of time elapsed between two consecutive lineups of the Sun, Earth, and Moon. In the upper part of this picture the three bodies are on a straight line; a synodic month later, the three are again aligned.

cause as the Moon goes around the Earth, the Earth continues to revolve around the Sun. The Moon must travel a little farther before the three bodies are in line again (Figure 2.16).

The Path of the Moon Around the Sun

With respect to the Earth, the Moon moves in a smooth ellipse with a speed that varies only slightly from an average value of 0.64 miles per second. With respect to the Sun, both the path and the velocity of the Moon are measurably different. The path of the Moon is wavy and its velocity is variable.

Approximately half the time, the Moon is outside the Earth's orbit and is moving in the same direction as the Earth. The speed of the Moon with respect to the Sun is the *sum* of its own velocity and that of the Earth—about 19 miles per second.

The rest of the time, the Moon moves inside the Earth's orbit and is moving in the opposite direction relative to the Sun. The speed of the Moon, then, is the *differ-*

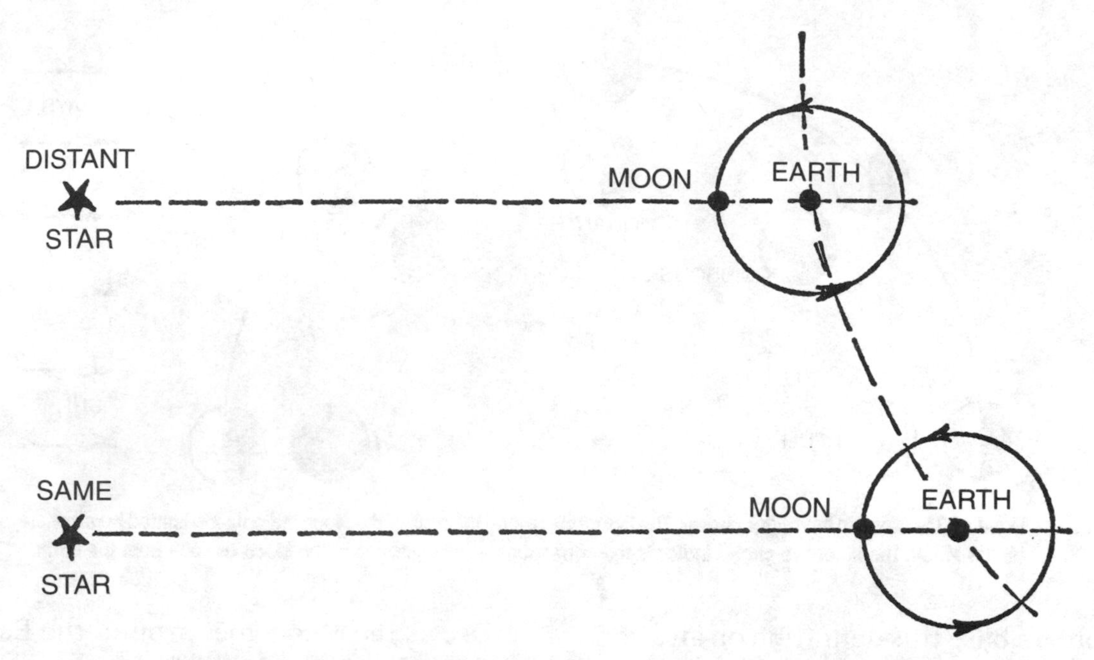

Fig. 2.16. The sidereal month. This is the period of one complete revolution of the Moon around the Earth, as seen from a distant star.

ence between its own velocity and that of the Earth—about 18 miles per second (Figure 2.17).

In reality (and to complicate things further), the path of the Earth around the Sun is also wavy in character and variable in speed. The center of gravity of the Earth-Moon system goes around the Sun in a smooth ellipse, the Earth and the Moon circling this common center of gravity. As a result, the Earth is at times inside the ellipse, and at other times, outside the ellipse following a wavy path at variable speeds.

The Period of Lunar Rotation

The Moon, while revolving around the Earth, is also rotating about its own axis. Its period of rotation is exactly equal to its period of revolution around the Earth. With respect to the fixed stars, the Moon completes one rotation in 27.32 days. The direction of the rotation, like the direction of the revolution, is counterclockwise as

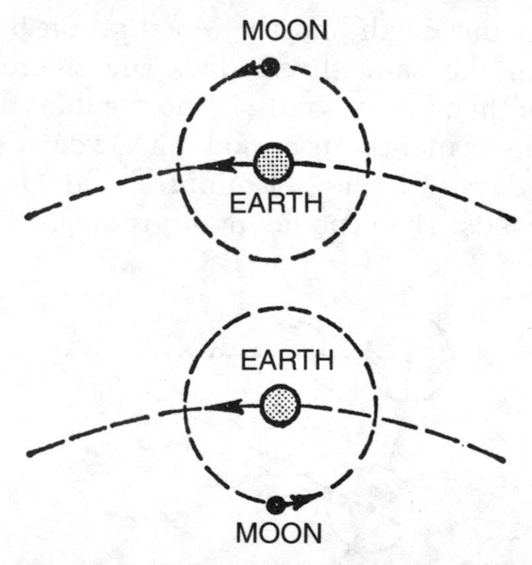

Fig. 2.17. The path of the Moon around the Earth as it orbits the Sun. Half the time the Moon is outside the Earth's orbit and moves in the same direction as the Earth; the rest of the time the Moon is inside the Earth's orbit and moves in a direction opposite to that of the Earth.

seen by an observer looking down from the Moon's North Pole.

The Earth's gravitational pull on a slight bulge on the side of the Moon facing the Earth is probably responsible for the equality of the rotation and revolution. It is also possible that this bulge of the lunar surface was originally formed by the Earth's gravity when the Moon was still in its plastic or liquid stage.

Since the two periods are equal, the same side of the Moon always faces the Earth, while the other side remains unseen. In fact, the far side of the Moon had remained unknown until it was photographed by Soviet and American lunar-orbiting satellites in the 1960s. If the speed of rotation and revolution were always the same, and if the Moon's axis were perpendicular to its orbit, then exactly 50 percent of the lunar surface would always face the Earth, and the other 50 percent would never be seen. Due to the effect of the Moon's slight rocking motions (librations), the ratios are slightly different. At most, 41 percent of the Moon always faces the Earth. Another 41 percent is never visible, and 18 percent is alternately visible and invisible.

Librations

These apparent to-and-fro motions of the part of the Moon facing the Earth are called librations. There are three kinds of librations: longitudinal, latitudinal, and diurnal. Most people can recognize the very familiar features on the lunar surface that make the man in the Moon. Longitudinal librations allow the observer to see not only the face of the Moon, but also the cheeks. Latitudinal librations allow the observer to see alternately the top of the forehead and the chin. Diurnal librations depend on the position of the observer on Earth.

Longitudinal librations of the Moon occur because the Moon is spinning on its axis at a constant speed, while its motion around the Earth is at a variable speed. The spinning is sometimes ahead of and sometimes behind the revolution, thus alternately exposing the left side and the right side. Latitudinal librations are due to the inclination of the lunar axis to the lunar orbit. The Moon's axis is tilted by 6½° to a line perpendicular to that orbit. Terrestrial observers can see 6½° past the northern pole of the Moon when that pole is tilted toward the Earth; two weeks later, observers can see 6½° past the South Pole of the Moon. Diurnal libration is a small effect, with a maximum of 1° difference in what the observer can see. It is due to the fact that two widely separated observers on the Earth would see slightly different hemispheres of the Moon.

The Lunar Surface

Even the unaided eye notices marks on the face of the Moon. While a telescope reveals the true nature of these marks, the closest investigations have been made directly from the lunar surface by several teams of Apollo astronauts beginning in 1969 and ending in the early 1970s. The surface of the Moon is covered with craters, maria (seas; singular, mare), mountains (including ridges and scarps), domes, rilles (crevices), and rays (streaks) (Figure 2.18).

A map of the moon. 1. Clavius; 2. Tycho; 3. Arzachel; 4. Alphonsus; 5. Ptolemaeus; 6. Albategnius; 7. Abulfeda; 8. Copernicus; 9. Eratosthenes; 10. Archimedes; 11. Autolycus; 12. Aristillus; 13. Cassini; 14. Plato; 15. Julius Caesar; 16. Flamsteed; 17. Kepler.

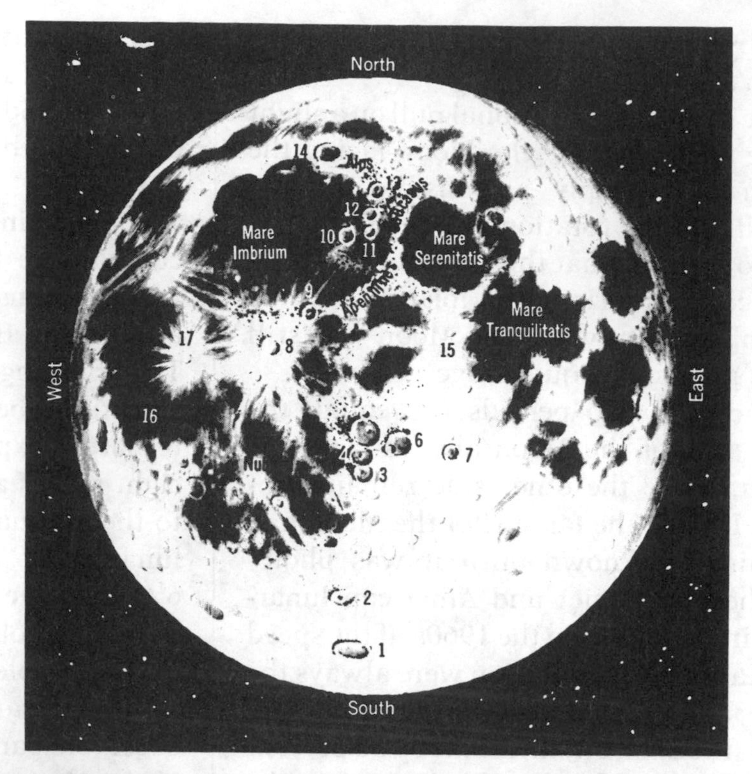

Fig. 2.18. Map of the Moon. (From Stuart J. Inglis, *Planets, Stars, and Galaxies*, 4th ed. New York: John Wiley & Sons, Inc. 1976, p. 101. Reprinted by permission of the author.)

1. *Craters* The outstanding feature of the Moon's surface are its many circular depressions, or craters. From the Earth, we can see some 30,000 craters with the aid of a telescope. The craters vary greatly in size, from about 150 miles in diameter down to fractions of a foot in diameter. It is estimated that there are perhaps 100,000 craters more than two miles across. Some of the largest craters visible on the near side of the Moon have been named for noted scientists and philosophers such as, Aristarchus, Clavius, Hipparchus, Plato, Copernicus, Kepler, and Tycho. The craters are found everywhere: In the three-quarters of the Moon's surface covered by rough, mountainous regions; in the quarter of the surface occupied by the maria; and, even craters within craters.

The rims of these craters also vary considerably in height. Some of the craters are bounded by high walls rising 3 or 4 miles above the surrounding terrain; others have rims only one or two feet high. The volume of soil piled up on the rim of each crater equals the empty space in the pit. This implies that something blasted the lunar surface material from the ground and deposited it around the impact site. It is widely believed that lunar craters resulted from tremendous meteoric bombardments when the Moon was still forming some 4 billion years ago.

2. *Maria* This name (Latin for "seas") was mistakenly given to the darker and smoother areas on the surface of the Moon by Galileo in the early seventeenth century. At one time they were indeed seas of

molten lava, but they have long since hardened to form a rigid crust. They differ from the rest of the lunar surface by being poor reflectors of sunlight; or, their **albedo** (reflectivity) is lower than the rougher terrain.

Many of the distinct maria have been named rather fancifully, considering the harsh reality of the lunar environment. Typical names are: Mare Crisium (Sea of Crises, or Conflicts), Mare Fecunditatis (Sea of Fertility), Mare Serenitatis (Sea of Serenity), and Mare Tranquillitatis (Sea of Tranquility, the site of the first Moon landing). The maria are roughly circular in shape and all but a few are interconnected. The diameters of the seas are in the hundreds of miles.

3. *Mountains* There are many mountain ranges, ridges, and scarps on the lunar surface, as well as many isolated mountain peaks. Several of the ranges have been named for terrestrial mountain ranges (e.g., Alps, Apennines); others for famous mathematicians and astronomers (e.g., Leibnitz). Some of the lunar mountains are much higher than terrestrial mountains. For example, several peaks in the Leibnitz Range are higher than Mt. Everest, the highest mountain on Earth.

4. *Domes* Many volcanic blisters dot the maria of the Moon, some measuring up to 6 miles in diameter. They resemble volcanic mountains on Earth such as the volcanic domes of northern California and may be remnants of partially filled or collapsed volcanic vents.

5. *Rilles* There are more than a thousand of these long crevices in the surface of the Moon, often found on the floors of craters. Some are narrow while others are more than a mile wide. They can be straight or winding and can cut across other surface features for distances up to hundreds of miles. The rille on the floor of the crater Alphonsus actually seems to be a line of smaller craters, perhaps resulting from the collapse of the surface material undermined by flowing lava. Another example is the Hadley Rille visited by Apollo 15 astronauts. The rille walls, cut into the base of the Apennine Mountains, displayed numerous layers of material—possible evidence of repeated lava flows.

6. *Rays* These are light-colored streaks on the surface of the Moon, that radiate in all directions from several of the more prominent craters, such as Tycho. Current opinion is that the rays are a result of surface material ejected when large meteoroids slammed into the lunar terrain. They are most visible during the full Moon.

Lunar Surface Gravity

The surface gravity of the Moon is only one-sixth that of the Earth. Thus the weight of an object on the surface of the Moon would be only one-sixth its weight on Earth. If you jumped on the Moon, you would rise six times higher than on Earth.

A direct result of the low surface gravity is a low escape velocity, which in turn accounts for the absence of any atmosphere. The escape velocity is 1.5 miles per second, so that a particle of gas having an initial velocity of 1.5 miles per second has enough speed to escape from the gravitational pull of the Moon. Since 1.5 miles per second is a sufficiently common value for atmospheric gases to move at temperatures on the Moon, if the Moon ever had an atmosphere, it has long since escaped.

Extensive lunar observations have verified the absence of an appreciable atmosphere. Experiments performed during a total solar eclipse show conclusively that rays of solar light grazing the surface of the Moon are not refracted. This finding indicates that there is no medium surrounding the Moon (in this case, atmospheric gases) that bends the rays of sunlight. In addition, the spectrum of light from the Moon is the same as that from the Sun, which also indicates that there are no gases on the lunar surface to produce any detectable changes in that spectrum. Upon studying many samples of lunar rocks brought back by the Apollo astronauts, researchers found no evidence that an atmosphere exists on the Moon. Initially, an atmosphere may have existed—but not for long. Some astronomers, however, are suggesting that a very thin atmosphere may actually exist on the Moon.

The Lunar Temperature

Temperatures on the Moon's surface vary greatly. The surface is continuously exposed to the rays of the Sun for a period of two weeks and then deprived of sunlight for an equal period of time. The difference in temperature between the light and the dark side is increased by the absence of an atmosphere and a low albedo.

Measurements taken on Earth of the sunlit side of the Moon register temperatures well above the boiling point of water (212° F). Other measurements taken of the dark side of the Moon indicate temperatures of about −225° F. The Sun's heat does not penetrate deeply below the Moon's surface. This is evident from studies of lunar eclipses. (Discussion of eclipses occurs later in this chapter.) The temperature at the Moon's surface drops rapidly as soon as the supply of sunlight ceases, with a change of around 100° F in one hour. The temperature rises with even greater speed soon after the surface emerges into the sunlight.

The Study of the Moon in the Space Age

The space age has added important technology to the study of the Moon. Before that age began, in the late 1950s, all our knowledge of the Moon was based on analyses of the visible, infrared, and radio waves that were reflected from the lunar surface. Some of the accomplishments in lunar exploration of the last several decades, presented chronologically, are as follows:

October 1959 The Soviet spacecraft Lunik 3 took the first photographs of the far side of the Moon, the surface not visible from the Earth. The far side turned out to have the same features as the near side, with two exceptions—it has fewer and smaller maria and many more craters.

1964 and 1965 The American Ranger spacecraft missions transmitted photographs as well as television pictures of the Moon, up to the moment of crash-landing on the Moon's surface. The photographs brought out a great deal of previously unseen detail on the lunar surface.

1966 to 1968 The American Surveyor series soft-landed a number of spacecraft on the Moon. The missions photographed and analyzed the lunar soil with the aid of special sampling devices. The main purpose of the missions was to gather useful information for the proposed manned landings. The analyses concluded the following:

- The lunar surface could support the weight of a vehicle.
- The soil was granular, and the grains varied in size.
- The soil particles stuck together, much like wet sand does on Earth.
- The soil could be compacted and had a density of about 1.5 grams per cubic centimeter.
- Many rocks were scattered on the surface; some of these were hard rocks, other crumbled easily when struck.
- The lunar soil material resembled crushed terrestrial basalt.
- The maria were primarily basaltic.
- The highlands were rich in aluminum and calcium.

1966 to 1968 The American Lunar Orbiter missions, originally designed to make a complete photographic survey of the Moon, were instrumental in discovering an increase in the gravitational pull over one of the large lunar seas in 1968. It is now assumed that there are large high-density mass concentrations called mascons under each of the large maria that causes the change in the gravitational pull.

1969 to 1972 Six American spacecraft with crews landed on the surface of the Moon. The first, Apollo 11, touched down near the Sea of Tranquility on July 16, 1969, and humans finally stepped onto another world. The mission ended with Apollo 17 in December of 1972. These expeditions determined the following:

- Dust is everywhere on the Moon, clinging to everything and everyone. It is usually mixed with small fragments of rock, forming a layer from three to twenty feet deep, called the regolith. Bedrock lies under the regolith.
- Some of the rocks scattered on the surface were probably formed by the cooling of lava some 4 billion years ago. Other rocks, called breccias, were formed by the cementing together of smaller, older rocks. In addition, fragments of iron meteorites and glassy particles were found on the lunar surface.
- The lunar rocks are very similar to the rocks formed by magma (molten rock) within our own Earth. The lunar rocks contain lower percentages of iron and higher percentages of the element titanium than their terrestrial counterparts; and they contain absolutely no water.
- Very sensitive seismographs, instruments used to record quake movement, were arranged on the lunar surface. Their readings indicated that the Moon is now mostly solid and has a relatively cool interior. The few moonquakes that triggered the instruments were probably caused by the tidal effect of the Earth (and occurred when the Moon was either nearest or farthest from the Earth), the impact of discarded spacecraft, or rare incoming meteoroids. Measurements of moonquakes have helped scientists to develop a model of the Moon that includes a crust, mantle, and core.

Formation of the Moon

The Moon's origin is still a mystery. Some believe the Moon formed in another part of the solar system and was eventually captured by the Earth's gravitational pull. Another theory suggests that a huge object struck the Earth, sending the debris into space, and eventually formed the Moon. Currently, the most widely accepted theory about the origin of the Moon suggests that it formed from the dust and gas from the early Solar System. Gradually, a ring of this material was orbiting the young Earth, attracted by the Earth's growing mass. Through a process of condensation, still not clearly understood, this ring of material became our Moon about 4.5 billion years ago.

After having condensing from this cold matter, the Moon (and the other members of the Solar System) underwent a period of intense bombardment from material left over from the formation of the Solar System. The rocks at the site of each collision had no time to cool before the next bombardment. The temperature of the Moon's surface rose dramatically as the meteorites rained down for millions of years. Eventually, the outer layer of the Moon melted, and remained molten until the bombardment stopped. The densest materials moved to the center and formed a small, partially molten core, while the least dense materials formed an outer layer. Materials of an intermediate density settled between the core and the outer layer, thus forming a mantle.

As the outer surface cooled, a thick, solid crust formed over the molten rock. Some of the meteorites that hit the surface broke through the crust, releasing molten rock through the breaks. As the molten rock from the interior filled the large basins on the surface, smooth maria were formed. Soon the Moon solidified and became a geologically quiet planet. As several hundred millions of years passed, a number of smaller meteorite impacts caused the severe cratering seen today in the lunar highlands. Occasionally, a huge meteorite, perhaps the size of an asteroid, would hit the Moon and blast out large craters.

As the number of meteorite bombardments diminished, the Moon's interior began to heat up as a result of the decay of radioactive elements below the surface. With the interior reaching the melting point, floods of lava would pour across the surface, having forced itself up through the shattered crust. These lava flows reflooded the large circular basins and created the maria we see today. Volcanic activity and lava flooding continued for about 800 million years and then stopped, leaving the Moon quiet once more. The last major lava flow seems to have happened about 3.1 billion years ago.

The Moon has been inactive ever since. The Moon, often called a small planet, probably lost its heat to space very rapidly. The warm, molten region was quickly replaced by a thickening layer of cold, strong rock. According to the Apollo data, the Moon is capped by a cold, 500-mile-thick zone of solid rock that would prevent any molten rock in the interior from forcing its way to the surface.

The absence of Earth-like crustal plates would prevent earthquakes, mountain-building, and any other of the familiar terrestrial surface-changing processes that have made our planet dynamic. In the time since the Moon became geologically dead,

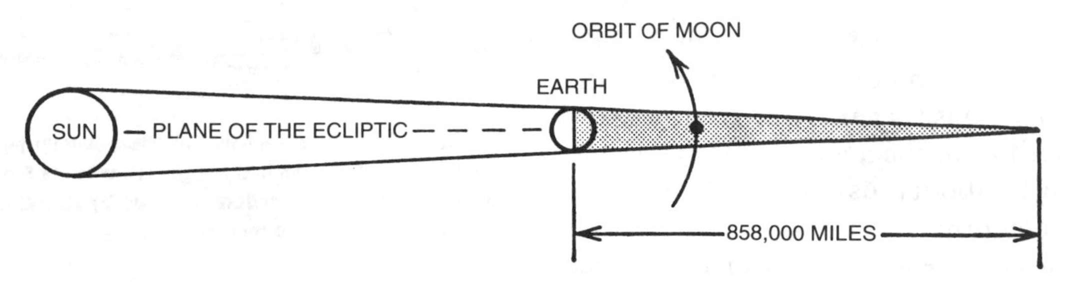

Fig. 2.19. Lunar eclipse. This kind of eclipse occurs when the Moon is inside the shadow-cone of the Earth. A lunar eclipse can be observed from any point on the night side of the Earth.

meteorite and micrometeorite collisions have moved only the top 50 feet of the lunar surface. This erosion rate represents one ten-thousandth of Earth's erosion rate. In reality, the Moon is one big fossil, and the very unchanging quality of the Moon's surface makes it an ideal laboratory for studying the early planet-forming era of our Solar System.

Life on the Moon

With its low surface gravity and high surface temperature, the Moon quickly lost any of its light gases, leaving no chance for an atmosphere to form. Therefore, the Moon is an arid and barren wasteland. There is no water, vegetation, or animal life. The absence of water also implies the absence of clouds. Without atmosphere there is no transmission of sound, hence no speech is possible. There is no twilight period and sunrises and sunsets are abrupt. Without an atmosphere, the sky appears ink-black. The Sun is a circle of brilliant light, while the Earth appears as a colorful sphere.

Eclipses of the Sun and the Moon

The Earth, in its orbital motion around the Sun, is accompanied by its shadow, which extends into space in a direction opposite to that of the Sun. The shadow has the shape of a cone; its base is the cross-section of the Earth, and its length is an average of 858,000 miles. The length of the shadow may differ from the average by about 25,000 miles because of the variation in distance between the Earth and Sun. A lunar eclipse occurs when the Moon enters the Earth's shadow-cone (Figure 2.19).

The Moon, following its path around the Earth, also carries its shadow along with

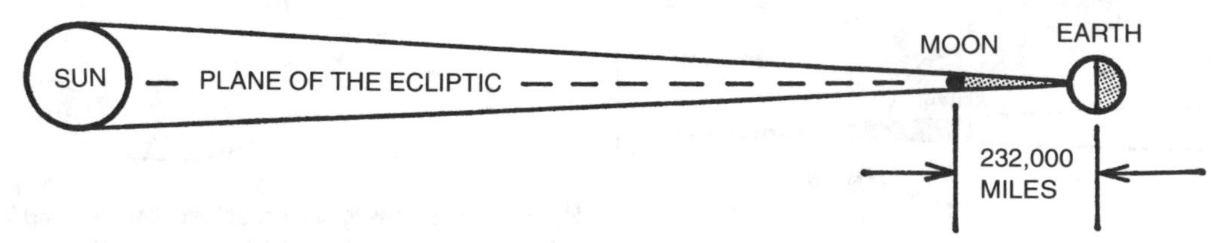

Fig. 2.20. Solar eclipse. A total eclipse of the Sun can be seen by terrestrial observers at all places on the Earth touched by the Moon's shadow-cone. When the shadow-cone does not quite reach the Earth, an annular (or ring) eclipse of the Sun takes place.

it. This shadow, too, has the shape of a cone, though much slimmer than the Earth's; its base is a mere 2,160 miles, and its length is on the average 232,000 miles, varying by about plus or minus 4,000 miles. A solar eclipse occurs when the Moon's shadow-cone reaches the Earth's surface (Figure 2.20). Often the Moon's shadow-cone is not quite long enough to reach the Earth because the distance between the Moon and the Earth varies from 226,000 miles at perigee to 252,000 miles at apogee. When the shadow-cone does not actually touch the Earth's surface, an event known as an annular or ring eclipse takes place. Under those conditions, the apparent cross section of the Moon is too small to cover the apparent diameter of the Sun, so the outer circle of the Sun remains visible as a brilliant ring.

A Lunar Eclipse

A top view of the ecliptic would mislead one to believe that there should be one lunar eclipse each month. This top view is shown in Figure 2.21. The apparent orbit of the Sun around the Earth (the ecliptic) is shown in this figure as the heavy line on the left, while the orbit of the Moon around the Earth is shown by the arrows on the right. The mistaken notion that there is a

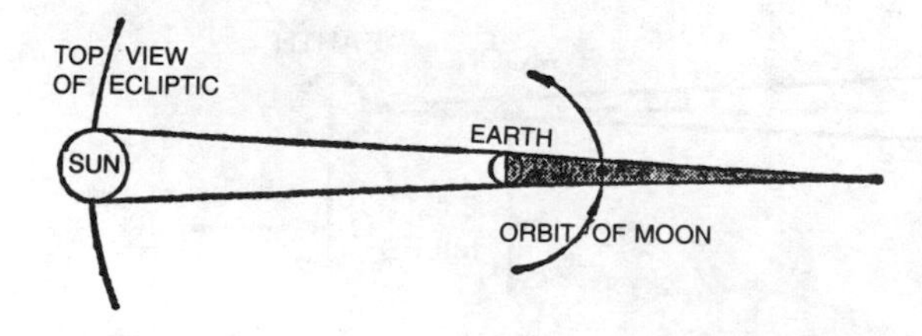

Fig. 2.21. Top view of ecliptic. Looking down at the apparent orbit of the Sun and the orbit of the Moon, one gets the mistaken impression that lunar eclipses should occur once every month.

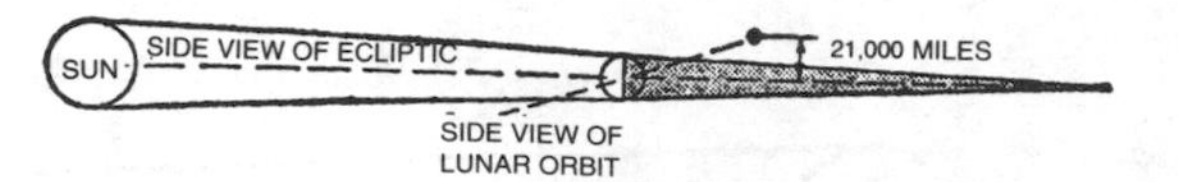

Fig. 2.22. Side view of the ecliptic. This view points up the fact that the Moon is actually not in a straight line with the Sun and the Earth. The Moon in its inclined orbit may pass by as much as 20,000 miles above or below the shadow-cone.

monthly lunar eclipse is revealed by the side-view, which shows that the three bodies are not in the same line in reality. Due to the slant of the lunar orbit to the ecliptic, the Moon may pass as much as 20,000 miles above or below the shadow-cone of the Earth (Figure 2.22). The Moon passes through the Earth's shadow fewer times than it passes above or below it.

For a lunar eclipse to occur, two important conditions must be met at the same time. First, the Sun, the Earth, and the Moon must lie on a straight line, as in Figure 2.23; that is, the Moon must be in its full phase as seen from the Earth. This happens once every month. Second, the

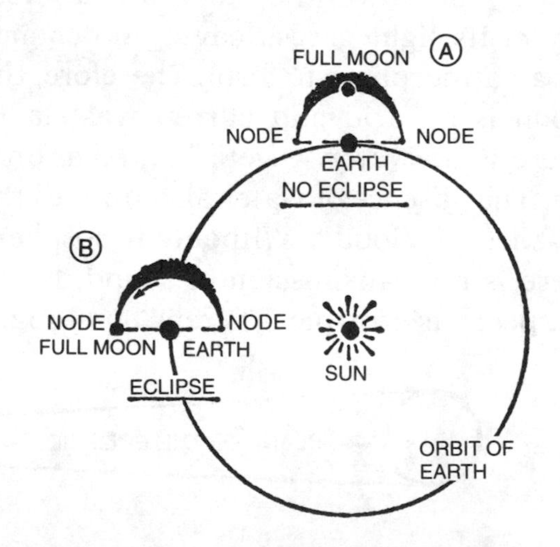

Fig. 2.23. Conditions for a lunar eclipse. Both at A and B the Moon is full. There is no eclipse at A because the Moon is well above the ecliptic; it will pass above the shadow-cone of the Earth. There is an eclipse at B because the Moon is in the plane of the ecliptic and will have to go through the Earth's shadow-cone.

Moon, moving in its orbit, should just be crossing the plane of the ecliptic; that is, it should be at one of the nodes.

It has already been observed that the Moon is below the plane of the ecliptic for half the month, and above it the other half of the month. The points at which the Moon crosses the plane of the ecliptic are known as nodes: the ascending node and the descending node. The line connecting the two nodes is known as the nodal line. Figure 2.23 shows two positions of the Moon in which the first condition is met. The Moon is full at both A and B. However, there is no eclipse at A, for the Moon is far above the ecliptic. There will be an eclipse at B because the full Moon is at a node. Both conditions can coincide as little as twice but no more than four times each year.

A lunar eclipse lasts a relatively long time because the Earth's shadow-cone (where the Moon crosses it) is nearly 5,700 miles wide. Because the Moon's diameter is 2,160 miles and its average speed is 2,000 miles per hour, if the Moon passes through the center of the shadow-cone, the total eclipse will last close to two hours.

The Earth's shadow does not hide the Moon completely. Even when totally eclipsed, it is visible. Its usual brightness is replaced by a dull, reddish color. This is primarily due to the refraction of sunlight by the Earth's atmosphere into the shadow-cone.

In the case of a partial eclipse, only part of the Moon passes through the Earth's shadow-cone. The normally full Moon then appears with a darkened notch either at its north or south end.

A Solar Eclipse

Eclipses of the Sun differ in several important ways from those of the Moon. First, solar eclipses can happen only at new Moon; lunar eclipses, only at full Moon. Second, all lunar eclipses, both partial or total, are visible from every point on the Earth that is facing the Moon. In the case of a solar eclipse, only the thinnest part of the Moon's shadow-cone ever touches the Earth. The maximum diameter of the circle intercepted by the Earth's surface is less than 170 miles. A much larger diameter is intercepted in the case of the penumbra—close to 4,000 miles.

The shadow-cone of the Moon is often called the **umbra** (Latin for "shadow"); the lighter region around the umbra is known as the **penumbra** (Figure 2.24). Observers in the penumbra region will see only a partial eclipse of the Sun, with the percentage

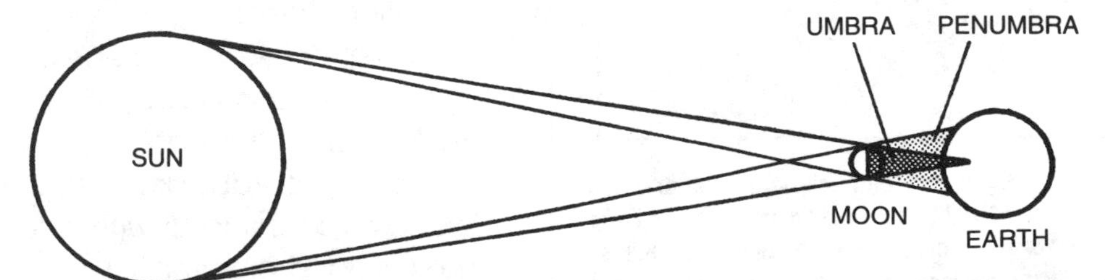

Fig. 2.24. The umbra and penumbra. The dark cone is the umbra. Observers located at that place on the Earth will see a total eclipse of the Sun. The half-dark region next to the shadow-cone is the penumbra. Terrestrial observers located there will see a partial eclipse of the Sun.

of the Sun's surface eclipsed depending on the distance to the umbra. The closer an observer is to the umbra, the more the Sun will appear eclipsed.

As the Moon and its shadow-cone move in their assigned orbits, the small circle and the circle due to the penumbra follow. A typical route taken by an eclipse is shown in Figure 2.25. The Moon's shadow-cone first touches the Earth near the west coast of South America and, moving eastward, leaves the Earth 3¼ hours later near the east coast of Africa. The width of the band of totality in this case is only about 100 miles.

The speed of the shadow on the Earth depends greatly on latitude and the angle the shadow-cone makes with the Earth's surface. At the equator, the speed may be only 1,000 miles per hour. In higher latitudes, especially near sunrise or sunset when the shadow-cone is greatly slanted to the Earth's surface, it may be 5,000 miles per hour.

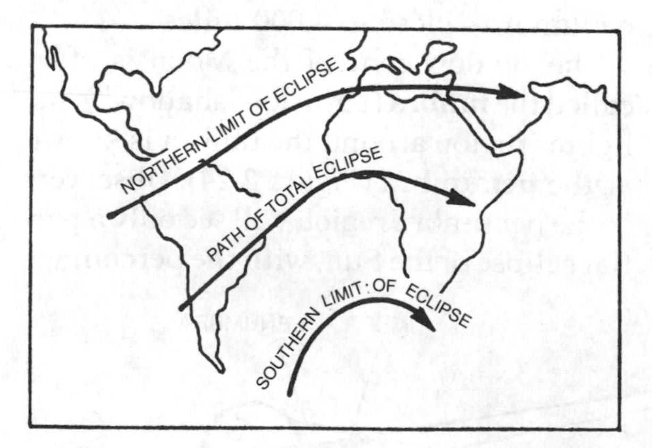

Fig. 2.25. Path of a typical total solar eclipse. The width of the totality path is less than 100 miles. The northern and southern limits indicate the regions of the Earth from which a partial solar eclipse can be observed. The arrows indicate the direction of the eclipse's path. (Reprinted from *The American Ephemeris and Nautical Almanac*, U.S. Naval Observatory, Washington, D.C.)

The duration of the total part of a lunar eclipse is nearly two hours. The greatest possible duration of a solar eclipse at any one point on the Earth's surface is 7 minutes and 30 seconds.

Total solar eclipses have meant many things to many people. To the primitive and superstitious, it has often caused great fear and concern. Battles have been suspended and peace treaties signed as a result of solar eclipses. For most of us lucky enough to witness a solar eclipse, it is simply a magnificent spectacle. The scientist is additionally interested because several important observations can be made only during the few minutes of totality. He or she may travel halfway around the world, and often to remote locations, in order to observe this event.

The silhouette of the Moon moves across the face of the Sun from west to east, covering more and more of the Sun's western limb. Several important stages may be observed (Figure 2.26).

WARNING!: Only observe the Sun with an approved solar filter. It is only during the brief moment of totality that one can look at the Sun without protection. Smoked glass or photographic film will not protect you from harmful ultraviolet radiation! *Permanent blindness can occur!* An effective way to observe the Sun and/or an eclipse is by projecting the Sun's image through a pinhole in a piece of cardboard onto another piece of cardboard; with a telescope, use a solar projection screen. This is the indirect method.

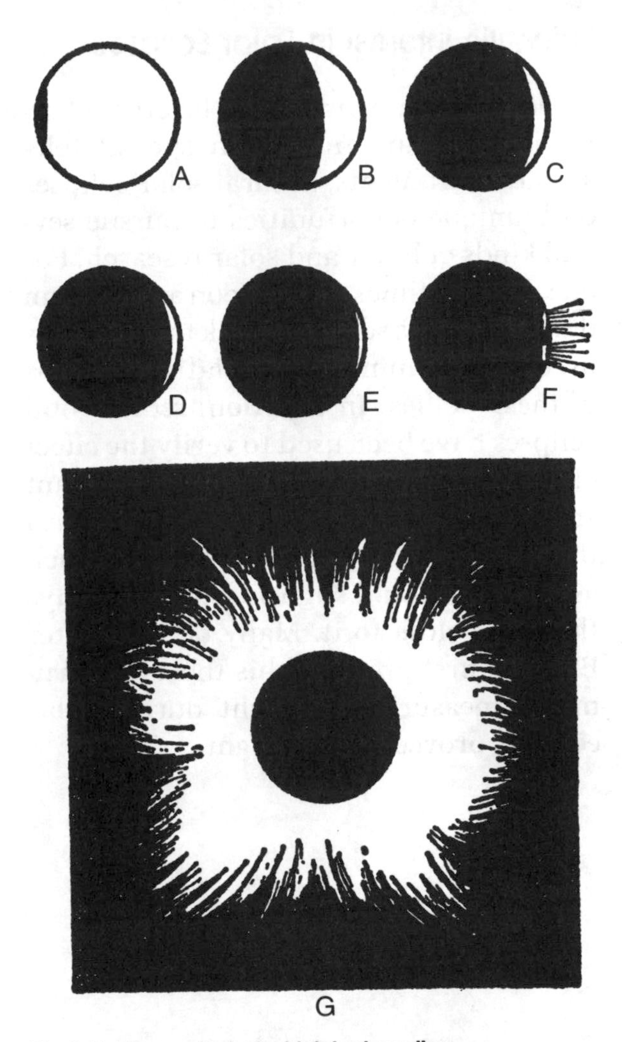

Fig. 2.26. Phases of a typical total solar eclipse.

1. At first contact, a small bite appears in the Sun's western limb. See Figure 2.26A.

2. As the darkened western limb becomes larger, both the intensity and the quality of the sunlight change—it has less blue than the light from the center of the disk. See Figure 2.26B.

3. In the last stages of the partial phase, the weird color of sunlight coming from the crescent is heavily accentuated. The dim, strange light seems to affect both animals and plants. Birds fly about, twittering; roosters crow; and, dogs bark with excitement. See Figure 2.26C.

4. Shortly before totality, hens go to roost, and many flowers close their blossoms as they normally do at sundown. Images of the crescent Sun may appear in the shadows of tree leaves which can act like many pinhole cameras. See Figure 2.26D.

5. Several minutes before the beginning of totality, ghostly shadow bands appear to be crossing over any exposed white surface. These bands are atmospheric waves made visible by the narrow crescent of sunlight. See Figure 2.26E.

6. A few seconds before totality, only several beams of sunlight reach the Earth through the valleys on the Moon's limb. These are known as "Baily's Beads." These brilliant beads of light vanish almost at once, and their disappearance marks the beginning of totality. See Figure 2.26F.

7. At totality, the full beauty of the solar corona is on display. This pearly halo surrounds the Sun, and often clearly defined streamers are seen flowing from the corona. Stars and planets make their appearance, greatly adding to the majesty of the scene. See Figure 2.26G.

Totality may last as much as 7½ minutes. The uncovering of the Sun then begins with the appearance of Baily's Beads on its western limb. All the phenomena that were seen at the eclipsing stage are repeated now in reverse order.

Catalogue of Eclipses

The Austrian astronomer Theodor Oppolzer (1841–1886) published a catalogue in which are detailed descriptions of nearly 8,000 solar eclipses and 5,200 lunar eclipses between 1207 B.C. and A.D. 2162. The tracks of all the solar eclipses are shown on the catalogue's nearly 160 charts. Today, of course, astronomers with the aid of computers have predicted eclipses to the year A.D. 4000. (See Appendix C for some of these future events.) The next solar eclipse visible from the continental United States will occur on August 17, 2017. The time of each eclipse can be predicted to within less than two seconds, and its path to within less than a quarter of a mile, computed on the basis of the positions and the motions of the Moon and the Sun. In the United States, these calculations are made by the Naval Observatory, Washington D.C.

Scientific Interest in Solar Eclipses

Although solar eclipses can be created artificially at any time with special telescopic instruments, natural solar eclipses offer unique opportunities to pursue several kinds of lunar and solar research. For one, contact times of the Moon and the Sun during eclipses serve to check the formulas used in determining the relative motions of these bodies. In addition, recent total eclipses have been used to verify the effect of the Sun's gravity on the path of distant starlight. According to Einstein's general theory of relativity, a massive object such as the Sun should slightly bend light rays that pass close to it. Many decades after Einstein first proposed his theory, instruments measuring starlight during solar eclipses proved he was right.

CHAPTER THREE

Tools of the Astronomer

KEY TERMS FOR THIS CHAPTER

aberration
altitude
angstrom
aperture
arc minute
azimuth
configuration
declination
dispersion
electromagnetic radiation
focal length
focus
focal ratio (f/)
refraction
resolution
spectrum

Telescopes

Much that is known about celestial bodies has been derived from the very small amounts of light that reach us from the depths of space. Detailed analyses of visible light have long been the mainstay of astronomy, supplying such information as chemical composition, temperature, location, and motions of celestial bodies. In recent decades, other forms of **electromagnetic radiation** (of which visible light is but one small window) have supplied the astronomical community with information undreamed of in the past. We now measure gamma rays, X-rays, ultraviolet light, infrared radiation, and radio waves from a host of familiar and exotic structures in the sky. Astronomers use a variety of instruments, both on the ground and in space, to enhance their understanding of the universe. However, the optical telescope remains the primary tool for the ground-based astronomer studying objects in visible light.

The telescope aids the astronomer in three distinct ways: It gathers the light from a celestial object, thus making the object appear brighter; it brings out details, such as small features on the Moon or Mars; and it magnifies or enlarges the image of the object that the telescope is examining.

There are two major types of telescopes: the refractor, which uses a spherical lens or combination of lenses to **focus,** or bring together the rays of light to one common point; and, the reflector, which uses a mirror to accomplish the same function. There is also a third type of telescope, called the catadioptric, that uses a combination of lenses and mirrors and is very popular with amateur astronomers because it is small and portable. We will discuss both the refractor and reflector types of telescopes in some detail, but first we will briefly examine how the human eye works, as it uses a lens much like a simple refracting telescope.

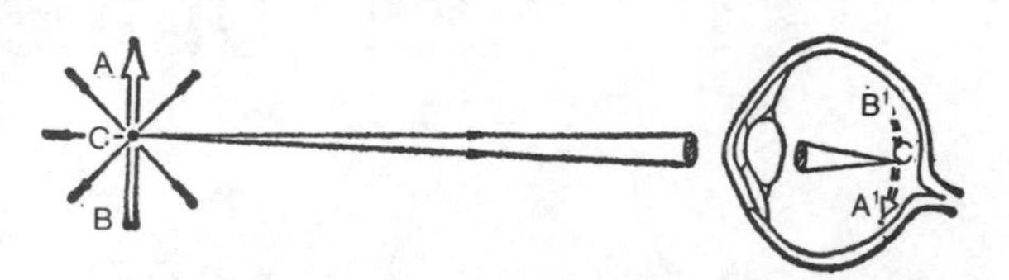

Fig. 3.1. The process of seeing. AB is a luminous arrow, every point of which is a minute source of light.

The light emanating from one of these points (point C) goes off in all directions (seven rays of light are indicated). Some of that light enters the observer's eye. The entering rays form a cone. The crystalline lens in the eye converges all the rays to one point C^1 on the retina. C^1 is the image on the retina of point C on the object. Every other point on the arrow will form a similar image on the retina.

The sum total of all the points on the retina produces the complete image A^1B^1.

The Human Eye

The human eye is a remarkable optical system, sensitive to visible light. It is like a simple telescope. The eye collects light through a small opening (the pupil), and, with a lens, focuses this light onto a light-sensitive lining (the retina). Humans see as a result of the light falling on the retina; the image formed in this way is changed into electrical signals and then sent to the brain via the optic nerve. The visual image formed on the retina has a one-to-one relationship with the object observed. For example, if one is observing the flame of a candle, every point of the flame must illuminate one, and only one, point on the retina in order to produce an undistorted image that can be understood by the brain.

Let us clarify this process even further. Let the object under observation be a luminous arrow, AB (Figure 3.1). The light emitted by a given point of this arrow, say point C, goes off in all directions in space. A small part of that light enters the observer's eye in the form of a cone.

To produce clear vision, all the light rays in the cone must come together to a single point in the retina. This task is performed by the crystalline lens in the eye, located just inside the pupil. C^1, the point on the retina is the image of the point C on the object. Images of all the other points of the object are formed on the retina in a similar way, collecting on the retina to form the image of the luminous arrow, A^1B^1.

The lens in the eye makes use of the curvatures of its surfaces to focus light. To see how this is done, let us follow two of these rays on their route from point C to point C^1. Ray 1, coming from point C, is refracted at the front surface of the lens, passes through the lens and is refracted a second time on crossing the rear surface of the lens. It finally falls on the retina at point C^1.

This **refraction,** or bending, is similar to the refraction that a ray of light undergoes in passing from air to water, from

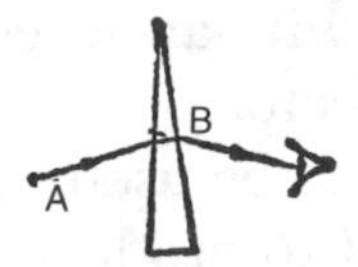

Fig. 3.2. Refraction of a light ray on going through a thin prism. The direction of the ray is changed on going from air (A) into glass. The direction is changed again on going from glass into air (B).

water into air, or in passing through a glass prism (Figure 3.2).

Ray 2, coming from point C, undergoes a similar experience: It is refracted at the front surface of the lens, then it passes through the lens. It is refracted a second time on crossing the rear surface. To obtain a clear image of point C, ray 2 must intersect ray 1 at the retina. The lens in the eye naturally adjusts its curvature to assure that the two rays intersect at the precise point.

Similarly, all the other rays coming from point C of the object, and entering the eye through the pupil, meet at point C[1] (Figure 3.3).

The glass lenses used in telescopes work in much the same way as the lens in the human eye. There is, however, an important difference between the two. The eye's lens is flexible and able to change the curvature of its surfaces, as small muscles squeeze or relax it. This allows the eye to focus rapidly on near and distant objects. In the case of a glass lens, the focus is fixed

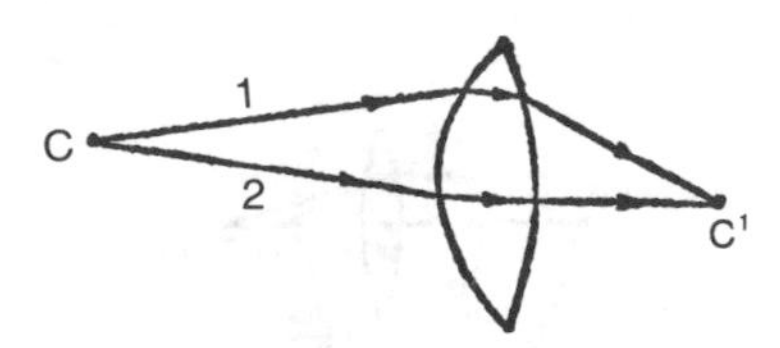

Fig. 3.3. The lens. All the rays reaching the lens from point C are refracted by the lens to converge at point C'. Each ray, such as 1 or 2, is refracted both on entering and on leaving it.

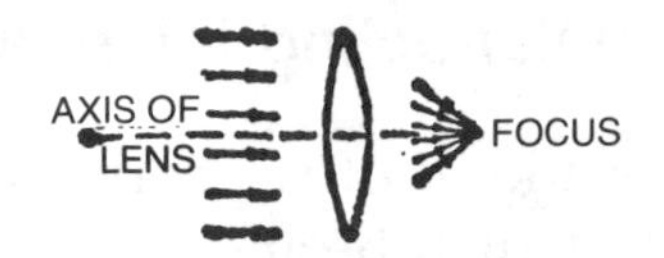

Fig. 3.4. The focus of a lens. The focus is a point on the axis of a lens where all the rays parallel to the axis meet. (Only part of the paths of the rays is shown.)

because the curvature of the lens does not change (Figure 3.4).

The Properties of Simple Lenses

Any lens is described or specified by indicating its **aperture, focal length,** and **focal ratio** (or f number, noted as f/). The aperture is the clear diameter of the lens through which light can pass; or, it is the diameter of the primary mirror of a reflecting telescope. Aperture is commonly used to describe the size of a telescope, such as the 40-inch Yerkes refractor, the 200-inch Hale telescope, or the amateur astronomer's 8-inch reflector.

The focal length is the distance between the center of the lens and the point where all the parallel light rays are bent to a common point or focus. The size of an image varies with the focal length. It increases as the focal length increases. Therefore, a lens with a focal length of 70 cm (centimeters) will form an image larger than a lens whose focal length is 60 cm. These three quantities are related in the following way:

f/ = focal length / aperture

Therefore, if a telescope has a focal length of 70 cm (28 inches) and an aperture of 15 cm (6 inches), its focal ratio is f/4.7.

The Simple Refracting Telescope

The simplest kind of refracting telescope consists of a long tube and two lenses. This is the familiar telescope often seen in many department stores, and it was the simple refractor that Galileo used to view the Moon and Jupiter in the early seventeenth century. The large lens that collects the incoming light is called the objective. Its function is to produce an image of the object at the focus or focal point of the telescope. The other lens, through which the observer views and magnifies the image, is called the eyepiece. If the object being observed is a point source, such as a star, the image formed at the focus will also be a point source. However, extended objects, such as moons, planets, nebulae, and galaxies, will form images that are upside down as well as reversed. According to the physical properties of lenses, light rays diverge (spread out) once again after reaching the point of focus, thereby creating an upside down image. In addition, any optical system having an even number of lenses or mirrors will produce a mirror, or reversed image (Figure 3.5). Therefore, simple refracting telescopes are not well-suited for terrestrial observation unless they are fitted with an accessory known as an image corrector.

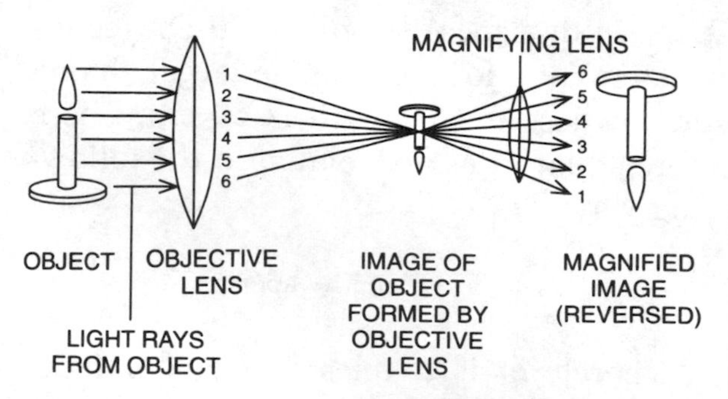

Fig. 3.5. Any optical system with an even number of lenses or mirrors will produce mirrored or reversed images.

In addition, to be useful at higher magnifications and to produce better images, the objective lens must remove two common optical defects. These defects are known as **aberrations** and are usually present in simple lenses. One of these flaws is known as chromatic aberration and the other as spherical aberration. The eyepiece has to be of a more complex design, too.

Chromatic or Color Aberration

A ray of ordinary, or white, light is actually a well-mixed combination of all the colors of the **spectrum,** the familiar rainbow of red, orange, yellow, green, blue, indigo, and violet. ("White light" signifies the ordinary light given off by the sun and the stars, and the reflected light of the Moon and planets.) Each color in this ray of white light refracts at a slightly different angle (called the index of refraction). Upon passing through a lens, the ray is not only refracted but also dispersed, or separated, into its component colors. This unmixing of the colors is a highly undesirable feature of a lens, producing a small rainbow of colors that forms a halo of unwanted color around the image (Figure 3.6).

This defect of the simple lens, by which the different component colors fail to arrive at the same focus, is called chromatic

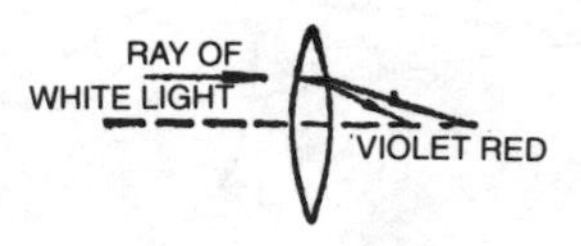

Fig. 3.6. Chromatic aberration. The violet component is refracted most, while the red component of the white light is bent least. All the other colors are intermediate between these extremes.

aberration. To reduce the effects of chromatic aberration, a compound lens is used, consisting of two different kinds of glass cemented together to form one unit (Figure 3.7). One part of the lens is a convex lens, with the center of the lens thicker than the edges. Its function is to bring the rays of light closer to the focal point. The other part of the lens is a concave lens, with the edges thicker than the center. Its function is to spread apart the light rays. Traditionally, the convex lens has been made from crown glass, which refracts light while slightly dispersing it into colors. The concave lens has been made from flint glass, which disperses light rather than refracts it. The concave lens offsets the refraction of the convex lens but does not totally eliminate it.

Today's manufacturers often use fluorite element glasses in their refractor objectives and have virtually eliminated chromatic aberration. This compound lens is known as an achromatic or color-free lens. Any given compound lens is actually color free for only two colors, such as green and red, or blue and violet. The two colors chosen are determined largely by the use of the telescope. Green and red are best suited for visual use, while blue and violet are best for photographic applications.

Spherical Aberration

A geometrical defect is produced when the surfaces of spherical lenses have not been properly ground or polished. Much to the dismay of the astronomical community, the mirror of the long-awaited Hubble Space Telescope launched in April of 1990, was discovered to be suffering from spherical aberration. Light rays cannot be properly focused by a lens or mirror with this defect. The light rays passing close to the edges of such a lens are refracted more than the light rays passing through the center. Again, the different parts of the observed object are focused at different distances from the lens and the result is a blurred image (Figure 3.8).

Fig. 3.7. A compound lens. Such a lens consists of two (or more) components cemented together or airspaced. Chromatic aberration can be greatly reduced by a proper choice of the quality of glass for each component.

Spherical aberration may be present even when there is no chromatic flaw. It can be corrected by making the face of the lens or mirror parabolic rather than spherical. A parabolic form is curved less at the edges than at the center and brings all parallel rays to a single, sharply defined focus. In fact, this correction is planned for the Hubble Space Telescope by the mid-1990s.

Correction of Aberrations

Achromatic objectives can be designed to correct both spherical and chromatic aberrations. The components of such a lens

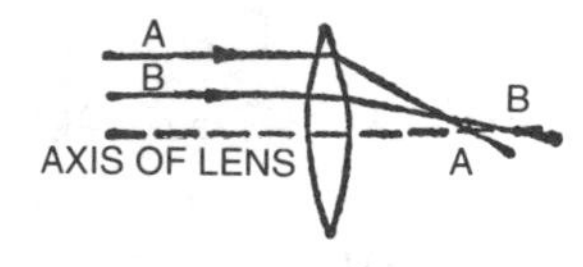

Fig. 3.8. Spherical aberration. Light ray A, close to the edge of the lens, is refracted more than light ray B. Parallel rays thus do not converge to one focus. This defect has nothing to do with the color of light. Spherical aberration is present even when the light is monochromatic.

need not have parabolic surfaces but instead can use ones that are spherical. There are two steps in designing such a lens: To begin, the lens maker must choose the type of glass needed for each component of the achromatic lens, such as one containing calcium fluoride. Then the proper combination of curvatures are calculated to eliminate the spherical aberration. This compound lens has no spherical aberration and is also corrected for chromatic aberration. However, this design usually requires long focal-length systems that make telescope tubes too long and unwieldy.

As a telescope objective, the three-element apochromatic lens is far superior to the ordinary two-element achromatic lens. In an apochromatic lens, three or more colors are focused simultaneously, and the images appear very sharp without any chromatic aberration. Therefore, the telescope designer need not resort to a long focal-length system, and the length of the telescope tube can remain relatively short and easy to handle.

Optical Coatings for Lenses

Only part of the light entering the telescope is actually transmitted through the lenses. Annoyingly, a sizable fraction is reflected from each optical surface. These reflections from curved surfaces produce troublesome secondary images (ghosts) and can substantially reduce the brilliance and clarity of the image. In recent years, telescope makers have reduced this shortcoming by coating each glass surface with a thin, transparent film, usually made of magnesium fluoride. This coating produces interference between light waves reflected at the top of the coating and the light waves reflected at the bottom of the coating. This interference effectively eliminates the ghost images and may increase the brightness and contrast of the image by as much as 30 percent.

The Reflecting Telescope

Isaac Newton built the first reflecting telescope in 1668. In the reflector, the function of the objective is performed by a highly polished, aluminum-coated mirror instead of a lens. (More expensive silver was once the standard coating but it tarnished too quickly.) The incoming light is brought to a focus by a concave mirror rather than a lens. The image formed by the mirror is viewed with an eyepiece, just as with a refracting telescope. Almost everything that was said about the refractor applies here. Only the optical **configuration,** or the way the parts of the system are put together, differs.

Unlike the ordinary household mirror, the aluminum on a telescopic mirror is put on the front, or concave, side of the mirror. The glass merely acts as a support for the aluminum coating. Having the metal in front of the glass eliminates absorption of the light. The image remains bright and does not suffer from chromatic aberration. The disadvantage is that the aluminum eventually oxidizes and the mirror must be periodically recoated.

In a reflecting telescope, the mirror sits at the lower end of the tube. The reflected light forms the image in the middle of the incoming rays. To be able to view the image through the eyepiece, the image must be moved. Two types of reflecting telescopes that have traditionally been used were designed by Newton and a French contemporary named N. Cassegrain. In Newton's arrangement, the light rays reflected off the primary mirror at the bottom of the tube are intercepted by a small, flat mirror that redirects the rays through the side of the tube and to the eyepiece. In some cases, a reflecting prism takes the place of the small secondary mirror (Figure 3.9).

In a Cassegrainian arrangement, a convex mirror does the redirecting. The incoming rays of light are intercepted by a convex mirror and brought to a focus through an opening cut in the primary mirror. One of the advantages of this arrangement is the flexibility in the focal length of the primary mirror. A set of convex mirrors used with the primary objective offers a variety of focal lengths in one telescope (Figure 3.10)!

Some reflectors use both the Newtonian and Cassegrainian focuses. Inevitably, the small secondary mirror or prism cuts off some of the incoming light, but this loss is relatively small (usually less than 10 percent). Furthermore, the obstruction cannot be seen at the eyepiece and does not interfere with the quality of the image.

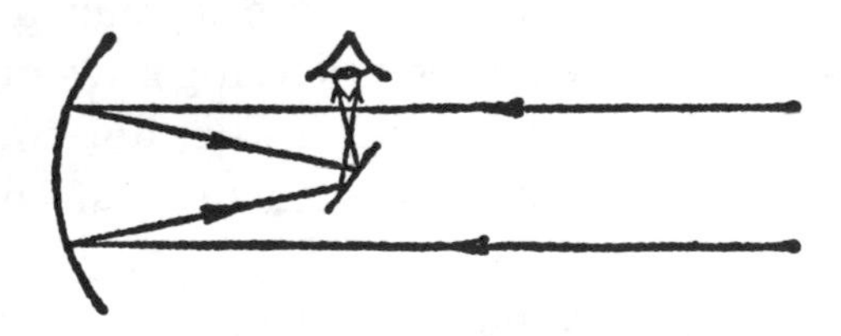

Fig. 3.9. The arrangement of the objective (mirror) and the eyepiece in the case of a reflecting telescope.

The mirror is at the lower end of the tube (the objective in a refractor, of course, is in the upper end).

The image produced by the mirror is in the middle of the incoming rays. An eyepiece could not be placed there as the observer would materially interfere with the incoming light.

In the Newtonian-type telescope, the one shown here, a little plane mirror diverts the rays through the side of the tube to the eyepiece.

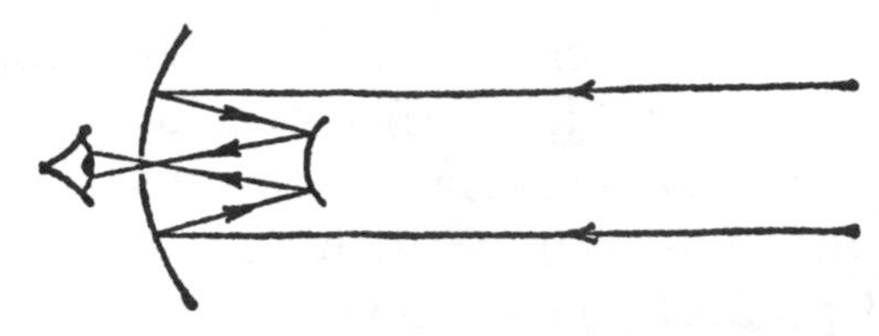

Fig. 3.10. The Cassegrainian arrangement. Diverting of the image produced by the objectives is done by a small concave mirror. The converging rays reflected by the objective are reflected by this concave mirror once more and brought to a focus just beyond an opening cut in the objective.

A particular reflector may be equipped with several convex mirrors having different curvatures. The focal length of the whole telescope is altered by changing the value of the curvature of the convex mirror.

Refractor Versus Reflector

Except for the primary objective and the design for routing the reflected light, there are no major differences between a refracting and a reflecting telescope. The various relationships between aperture, focal length, and focal ratio of each type of telescope are identical. Each telescope design has its advantages and its disadvantages.

Historically, the refracting telescope was the first kind invented and remained practical for astronomical research into the twentieth century. It offered excellent **resolution,** the ability to separate and see detail in an image, and enhanced contrast. A tube protected the interior from the elements. However, in order to optimize its performance, designers were soon forced to create lenses and tubes too large to be reasonably built. Therefore, while refracting telescopes have faded from professional research, they are popular (albeit in more portable sizes) with many amateurs who are especially interested in observing planets and splitting double stars.

The reflecting telescope has gained popularity in both the professional and amateur worlds of astronomy. The reflector is free from chromatic aberration; its possible shorter focal lengths allow for a much more compact design; the glass of the mirror need not be perfect, as the light is reflected from a coated surface; only one surface has to be shaped with exact precision; no light is lost by absorption when passing through glass lenses; and, it is much cheaper to produce.

What Telescopes Do

Telescopes have three functions:

1. They increase the apparent brightness of an object. This increase in brightness is the light-gathering power of the telescope.
2. They bring out detail that cannot be seen with the unaided eye. This is called the resolving power of the telescope.
3. They magnify the object, or make it appear that the object is closer. This is called the magnifying power of the telescope and is determined mainly on the choice of the eyepiece.

The most important function of a telescope is to gather a large quantity of light from a celestial object under observation. Simply stated, the larger a telescope's objective is, the greater its light-gathering power. Astronomers are constantly attempting to build bigger telescopes for this reason. The light-gathering power makes it possible to see stars and galaxies that are much too faint to be seen with the naked eye.

The resolving power of a telescope is intimately connected with the clarity with which details can be seen. The greater the resolving power, the clearer the detail will be. For example, a point of light that appears to the naked eye as one star may be separated (resolved) into two or more close stars when viewed through a telescope with high resolving power. This function is not to be confused with magnifying power defined below.

Consider two pinpoints of light, such as the headlights of a car. At a distance of several feet, the two headlights will appear as two distinct and separate sources of light. At a greater distance, the two will merge into one relatively fuzzy point of light. Experiment shows that the pinpoints of light can no longer be separated when the angle between them at the eye is less than about one **arc minute** (an arc minute is $\frac{1}{60}$ of a degree). In other words, the resolving power of the normal eye is about one arc minute. However, when viewing an object through larger and larger objective lenses, resolving power increases, thereby allowing a finer and finer separation of objects measured in arc seconds.

The magnifying power of a telescope depends on both the focal length of the objective lens and the focal length of the eyepiece. In order to compute the magnification factor of any combination of objective and eyepiece, all one has to do is divide the focal length of the objective by the focal length of the eyepiece. For example, if a telescope whose objective has a focal length of 2000mm (millimeters) is used with an eyepiece whose focal length is 20mm, then the magnifying power is 100x.

The formula for magnifying power seems to indicate that there is no upper limit to magnification. Why not produce a telescope whose magnifying power is one million? The laws of optics place strict limits on magnification. The resolving power of a telescope, for one, limits magnification. In addition, an increase in magnification causes a decrease in brightness. The movement of the Earth's atmosphere also will affect the view with increasing magnification. Stars will appear to twinkle more and planets will appear to be under water. Finally, increasing the magnification decreases the actual field of view of the sky.

The relationship between magnification and actual field of view is illustrated in Figure 3.11. In all three views, the telescope is pointed in the same direction. With low magnification, the tops of two buildings and a large portion of the sky were in the field of view. With higher magnification, only the tower of one of the buildings can be seen through the telescope. With still higher magnification, one window occupies the whole view. When an observer wishes to view an extended object such as a dim galaxy or nebula, the advantages of low magnification are obvious.

The above reasons dictate that there is a practical limit to maximum magnification. Experience has shown that 50 to 60 times for each inch of the diameter of the objective is all that is useful. For example, a 4 inch telescope could reach a maximum of 200x. There is also a minimum for useful magnification, usually 4 times for each inch of the objective's diameter. If the magnification is less than this, the column of light leaving the eyepiece will be too large to enter the pupil of the eye and some of the light will be wasted.

Other Telescope Designs

There have been many innovations in the field of telescope design in the twentieth century. In 1931, Bernhard Schmidt (1879–1935) invented a lens-mirror arrangement that made it possible to use a more easily produced spherical mirror. The aberration caused by the spherical shape of the mirror is corrected by a thin lens known as a "corrector plate" placed at the center of the mirror's curvature. At about the same time, the Russian optical designer, D.D. Maksutov, created a variation of the Schmidt design. He chose to use a thicker, less complicated meniscus lens with a spherical surface to provide the correction for the mirror (Figures 3.12 and 3.13).

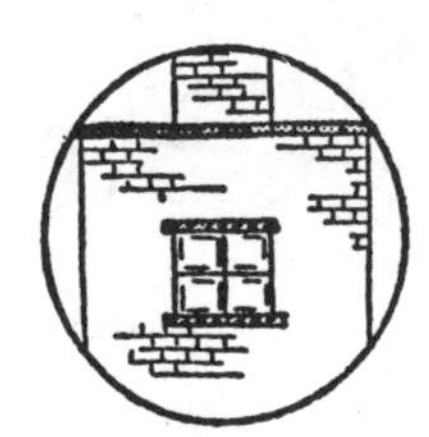

Fig. 3.11. Left: Low magnification. Center: Moderate magnification. Right: High magnification.

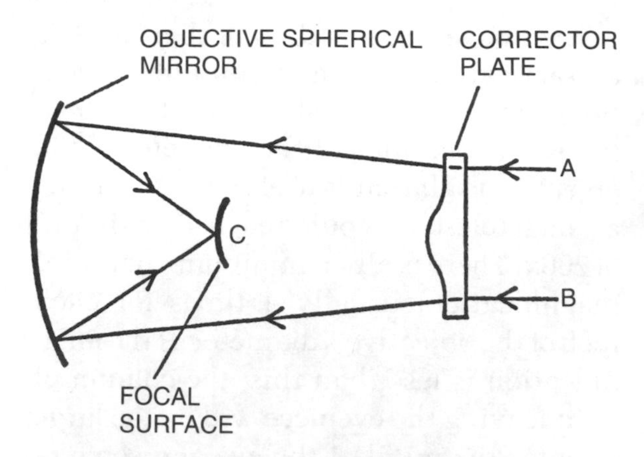

Fig. 3.12. Rays A and B are deflected slightly by the corrector plate, causing them to arrive at the same point C. Point C is also reached by rays close to the axis (not shown).

There are advantages to these two designs: It is easier to manufacture the spherical mirror, and because the lens-mirror arrangement actually folds the path of the light rays, a much shorter tube can be built. A telescope of a pure refractor or reflector design would need a tube several times the length in order to achieve the same focal length. For professionals, and more advanced amateurs, this has meant taking very fast astrographs, astronomical

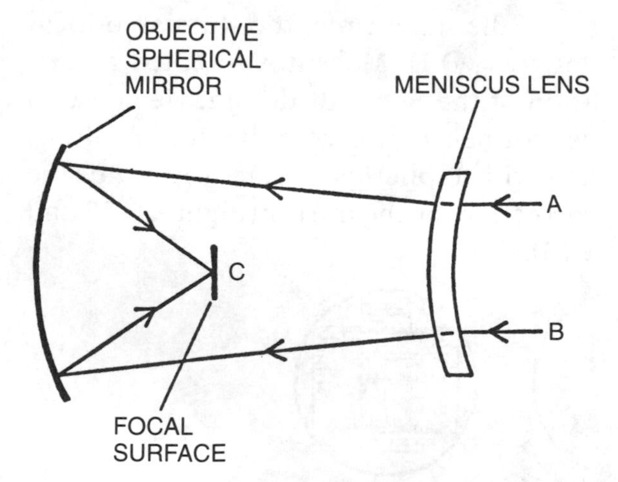

Fig. 3.13. The diverging of the rays is accomplished by the meniscus lens. The lens has spherical surfaces and thus is easy to manufacture.

photographs, used primarily to photograph large areas of the sky on each photographic plate. For example, the 48-inch, Palomar Observatory Schmidt telescope in California was very useful for the Palomar Sky Survey, yielding a complete catalogue of deep-sky objects in amazing detail. In the last 20 years, amateurs, too, have enjoyed the fruits of these newer designs. Very popular today are the portable and compact 8- to 14-inch Schmidt-Cassegrain telescopes. They are fairly easy to operate and allow amateurs to take quality astrophotographs.

Other telescope designs have emerged in recent years. Most noteworthy is the use of multiple mirrors and active optics in various professional observatories around the world. Multiple Mirror Telescopes (MMTs), as the name implies, use up to several dozen of the same-size mirrors to produce the results of a single mirror. Through the use of computerized electronic processing, all the images gathered by the mirrors of a MMT are combined into one final picture. In addition, small computer-controlled supports, called activators (thus the term "active" optics), constantly change the shape of the mirrors to counteract the downward pull of gravity. This computer control keeps the mirrors in perfect shape, otherwise the mirrors would sag beyond usefulness and produce worthless images.

Modern Detectors for Telescopes

For more than half of this century, most astronomical observations were recorded using photographic film or plates. Photography had many advantages over direct vision:

• Photographic plates detect stars that are many times dimmer than the faintest stars seen through the same telescope. Because the change in the chemicals of the plate is a cumulative effect, the image actually builds on the photographic emulsion. The human eye does not build up an image, as the light from the stars does not accumulate on the retina.

• A long-term exposure produces details usually unseen by visual observation. Much of our knowledge of remote galaxies comes from detailed photographic images.

• A permanent record is especially important in studying changes in brightness and the relative positions of celestial bodies. Some seemingly unimportant stars may suddenly become prominent; that is, they may explode as supernovae or suddenly increase in intensity. Existing photographic records can be examined to see if similar behavior occurred in the same region in the past.

• Sections of the sky can be studied at leisure. Some celestial bodies are only in the sky for short periods. Photographs capture such short-term events, allowing the astronomer to study the phenomena at his or her convenience.

• Photographs can be enlarged and studied with the aid of a microscope. Such magnification of photographic images is especially helpful to count stars, and is of particular interest for astronomers studying globular clusters.

• Photography has been used a great deal in the study of the Solar System. The planet Pluto was discovered in 1930 by Clyde Tombaugh (1907–) who studied thousands of photographic plates before the tiny planet was found. (Stars show up as single points of light on a photograph, while moving objects like planets, comets, and asteroids, leave short streaks.)

Today, however, electronic technology has almost replaced the use of photographic emulsions. The amateur astronomer still has a wide range of films and accessories available for recording observations, but the majority of professionals, and some advanced amateurs, are now using sophisticated devices such as the photon multiplier tube and the charged-couple device (CCD) to enhance and record their observations. These devices work by collecting the individual photons (electromagnetic energy in its particle form) of light and building a greatly enhanced image photon by photon. With the aid of a computer, images can be recorded, stored, and manipulated. The computer images can even be given different colors to emphasize whatever the astronomer wishes to study, such as the chemical make-up and temperature of an object. All this can be accomplished in a fraction of the time necessary with conventional photography. In fact, the electronic revolution has made it possible for astronomers to control telescopes from thousands of miles away by using telephone and satellite hookups. In some systems, the telescope is remotely controlled by a computer linked to an Earth-orbiting satellite then to another ground-based computer where astronomers monitor the data. In other systems, the computers are directly linked through telephone connections.

Telescope Mountings

One of the most important parts of a telescope is the mounting. Much engineering skill and ingenuity are devoted to the mounting. It not only holds the telescope and all its accesories, but it also allows the telescope to be pointed in any direction of the sky so that celestial objects can be tracked continuously. A well-designed mounting minimizes image-destroying vibrations, and moves smoothly and precisely.

The simplest mounting unit is a combination of vertical and horizontal axes. The telescope is attached to a fork through horizontal bearings, and thus can be rotated from horizon to zenith (the point in the sky directly overhead) through various **altitudes,** or degrees above the horizon. The fork, in turn, is able to rotate on a vertical axis through 360° of a horizontal circle, at every **azimuth** (horizontal angle) from 0° to 360°. This is called the altitude-azimuth mounting, or alt-az for short (Figure 3.14). However, it cannot easily be used for extended periods of observation or long-exposure photography, as the constantly rotating earth makes all objects in the sky appear to move in both altitude and azimuth. The alt-az mounted telescope would have to be continuously adjusted both horizontally and vertically. With their tremendous weights, many new professional telescopes must use an alt-az mounting, but they have the advantage of being automatically adjusted by sophisticated computer programs. For example, the largest single mirror telescope, the 236-inch reflector at the Zelenchukskaya Astrophysical Observatory in the Caucasus Mountains of the former U.S.S.R. (750 tons), uses an alt-az mounting and is controlled by a digital computer.

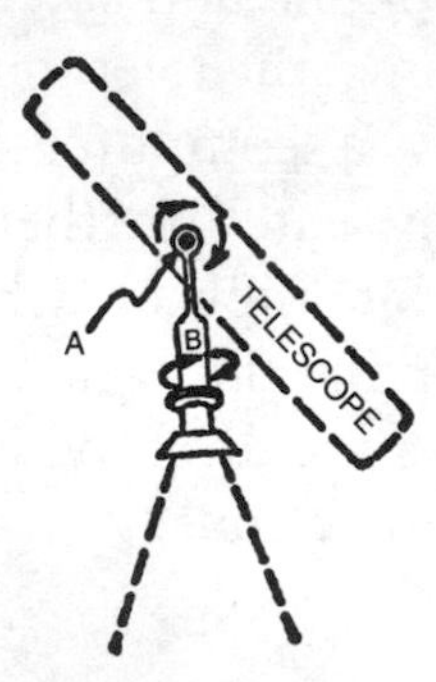

Fig. 3.14. Alt-az mounting. The telescope can be rotated around the horizontal axis A and can be directed to any altitude from the horizon to the zenith.

The telescope together with the horizontal axis can be rotated around the vertical axis B to any azimuth from 0° to 360°.

Another type of mounting is specifically designed for the purpose of keeping a celestial object in view for long periods of time. With an equatorial mount, only the angle corresponding to the apparent motion of the object from east to west, known as the object's right ascension, has to be adjusted. This adjustment is usually performed by a small electric motor with a set of precision gears called a motor drive. This mounting also has two axes at right angles to each other. One axis is called the polar axis and is parallel with the earth's polar axis. The other axis, known as the **declination** axis, rotates around the polar axis. Once an object's declination, or position above or below the celestial equator, is locked in, the motion of the object can be continuously tracked with only a small motor moving the telescope around the polar axis. Since the motor moves at the same rate as the turning earth, only in the opposite direction, the object appears to be fixed in the telescope's field of view (Figure 3.15).

Radio Telescopes

Stars emit not only visible radiation (light) but also shorter (X-rays) and longer (heat

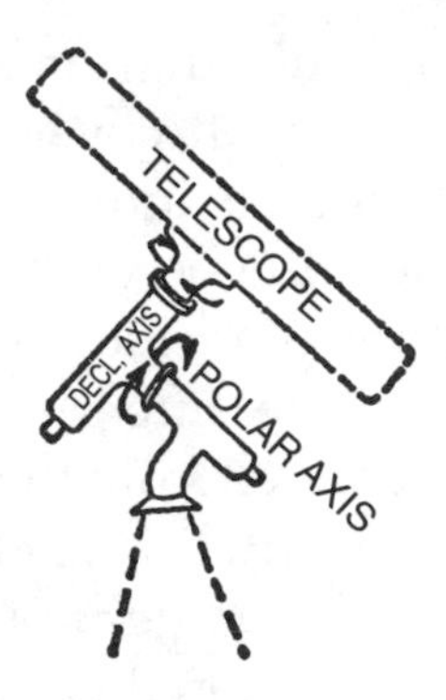

Fig. 3.15. The equatorial mounting. The telescope is rotated around the declination axis to the desired declination and clamped firmly at that position.

The telescope together with the declination axis rotates around the polar axis to keep the object continuously in the field of view.

and radio) wavelength radiation. In fact, stars, like all hot bodies, emit radiation in all parts of the electromagnetic spectrum. The Earth's atmosphere, however, is transparent only to light and radio waves. All other radiations given off by stars do not reach us, as they are absorbed for the most part by the Earth's atmosphere. The wavelengths of light that can get through the atmosphere (the optical window) range from about 4,000 angstroms (violet) to about 7,000 angstroms (red). The radio waves that penetrate the atmosphere (the radio window) range from about 1/100 of a meter to about 30 meters.

In 1931, an electrical engineer, Karl G. Jansky (1905–1950), accidently stumbled on the existence of radio waves from space while attempting to solve the annoying hiss experienced by trans-Atlantic telephone communications between London and New York. He eliminated every possible source of this interference until it was evident that the source was the center of our own Milky Way galaxy! His accidental discovery opened up the new science of radio astronomy.

The final product of an optical telescope is either a photograph or a direct visual observation; whereas the information obtained from the radio telescope is in the form of fluctuating electrical signals that are recorded on magnetic tape and analyzed by a computer. The radio waves reflected from a parabolic dish reach a receiver placed at the focus of the paraboloid. The waves are amplified and then sent to a controlling computer that stores the information for later use.

There are several similarities between a radio telescope and an optical reflecting telescope. Both have mirrors, usually parabolic in shape; both use equatorial mountings; and both are designed to have high resolving powers. There are also many differences. A reflecting telescope's optical mirror is a finely polished piece of glass coated with aluminum or silver. The radio mirror (often referred to as a "dish") is made of wire mesh or machined metal sheets. Another important difference is that a radio telescope can operate anytime. A radio telescope does not need dark skies to receive radio signals; and in inclement weather, radio waves can penetrate clouds. In addition, radio waves are not blocked by the huge amounts of interstellar dust and gas that fill large portions of the universe, making it easy to observe optically invisible objects in the radio spectrum. Lastly, radio telescopes do not have to be situated on mountains high above the atmosphere, or away from lights like optical telescopes. Radio observatories must be located far from radio and television signals, and from static emitted by car and airplane ignition systems.

It should be noted that the resolving power of a radio telescope is not as great as an optical telescope because radio waves are much longer than light waves. Several means to improve resolving power have been devised. As mentioned earlier, an increase in resolving power seems as simple as building larger and larger dishes. As with optical telescopes, there are practical limits to the size and cost of building a single, large radio dish. Instead of one huge dish, engineers have created the radio interferometer. It consists of two or more radio telescopes separated by a distance anywhere from several yards to thousands of miles. The separation acts as the effective diameter of the telescope. The radio waves are converted to electrical signals by each telescope and are combined by computer programs that match the signals to within one-millionth of a second of each other.

One such dish system is known as Very Long Baseline Interferometry (VLBI). Radio telescopes as far apart as South Africa and Australia have been combined to create a telescope with an effective diameter of 6,000 miles! A variation on this design is the Very Large Array (VLA). The VLA, near Socorro, New Mexico, consists of 27 radio dishes mounted on railroad tracks in the shape of the letter Y. The dishes can be moved around the tracks to create a radio telescope that is effectively 13 miles in diameter.

New Windows

Orbiting satellites high above the blanket of Earth's atmosphere have provided astronomers with new windows onto the universe. Observations made in the wavelengths of X-rays, gamma rays, ultraviolet light, and infrared radiation are now routine. Some major discoveries made in these wavelengths include: Neutron stars and black holes; the possible existence of antimatter in the universe; the abundance of elements in the interstellar medium never before detected from the ground; the interstellar heavy hydrogen element, deuterium that may shed light on the density of the universe; cool, red giant stars; and, the possibility of a solar system in the making around the star Vega.

A Tale of Two Telescopes

As astronomers strive for more detailed knowledge of the universe, they naturally look to larger and more sophisticated telescopes. The largest, single-mirror, optical telescope in the United States called the Hale telescope was completed in 1948 and sits on Palomar Mountain in California. The mirror of this telescope is 200 inches in diameter and weighs nearly 15 tons. Grinding and polishing this mirror took eleven years, producing a surface that is accurate to within one-millionth of an inch. Combined with its mount, the entire instrument weighs 500 tons but is so delicately balanced that it can be pointed in any direction of the sky by an electric motor a few inches in size.

The powers of the Hale telescope, nicknamed the Big Eye, are enormous. It gathers as much light as do a million human eyes. With its aid, one can see candlelight at a distance of 10,000 miles. It penetrates twice as far into space—a distance of 2,000 million light-years—than the 100-inch telescope on Mount Wilson. After it was built, this telescope quickly demonstrated that the previous yardstick for astronomical distances was incorrect. The distance to the Great Galaxy in Andromeda, thought to be 750,000 light-years, was determined to be 1,500,000 light-years in 1952, and later corrected to 2 million light-years. This finding implied that the universe was larger and older than previously believed. In the 1960s, the telescope was used to study the enormous red-shift in the unusually energetic and distant objects called quasars. (Quasars will be discussed in more detail in Chapter 13.) Although no longer at the cutting edge of research, the Big Eye continues to make observations deep into space and remains one of the engineering wonders of our century.

The story of the Hubble Space Telescope (HST) is one of triumph and disappointment. Lofted above the blurring effects of Earth's atmosphere by the Space Shuttle Discovery on April 24, 1990, its 94-inch primary mirror and array of sophisticated detectors were to provide astronomers with views of the universe never before achieved. Like the Hale Telescope on Palomar, the HST represented the newest in design and technology of the day, with enough power to probe the edge of the visible universe. Unfortunately, soon after a preliminary test of its instruments, a serious problem in the primary mirror was detected. The mirror suffers from spherical aberration, preventing any images from being sharply focused. On a future shuttle mission, astronauts will add corrective optics to the telescope. Meanwhile, scientists at the Space Telescope Science Institute, where the HST data is collected and stored, are continuing to gather other data from the HST. In many cases, even with all of its problems, the HST is outperforming ground-based observatories in such areas as photographing Mars and Saturn, observing stars in ultraviolet light, and probing the depths of globular clusters like M15, where HST's cameras revealed no indication of a suspected black hole.

Spectroscopy

Telescopes allow astronomers to view or record the appearance of objects in the sky, but it is through the science of spectroscopy that astronomers can actually analyze an object's nature. By breaking starlight into its component colors (a spectrum), astronomers can determine the surface temperature of any star; the chemical proportions that make up the star; any magnetic fields associated with the star; and the speed with which the star is either approaching us or receding from us. All this data, as well as other information, comes from careful analysis of the star's radiation. The branch of astronomy that deals with such studies is called spectroscopy and its basic tools are the spectroscope and the spectrograph.

The Spectroscope and the Spectrograph

A spectroscope disperses light into its component (primary) colors, forming a spectrum. This process is similar to how water droplets in the atmosphere disperse sunlight and create rainbows. A spectroscope uses either a glass prism or a diffraction grating to break up the light. When the spectrum is viewed directly with an eyepiece, the instrument is called a spectroscope. However, a more common practice is to replace the eyepiece with a camera or an electronic device that can make a permanent record of the spectrum. This instrument is called a spectrograph.

The Prism Spectroscope

A ray of ordinary or white light, such as sunlight, will be dispersed once upon entering the glass of a prism, and then once again upon leaving it, creating a continuous array of colors called a spectrum (Figure 3.16). Two basic physical laws govern the creation of a spectrum. First, light is a form of energy traveling in waves. Experimentally, red light differs from blue light only in the wavelength (or equivalent energy) associated with that color. Red light has the longest wavelength (lowest energy) in the visible spectrum whereas, violet has the shortest wavelength (highest energy). The wavelength is the horizontal distance between crests, or valleys, of two adjacent waves (Figure 3.17). This distance

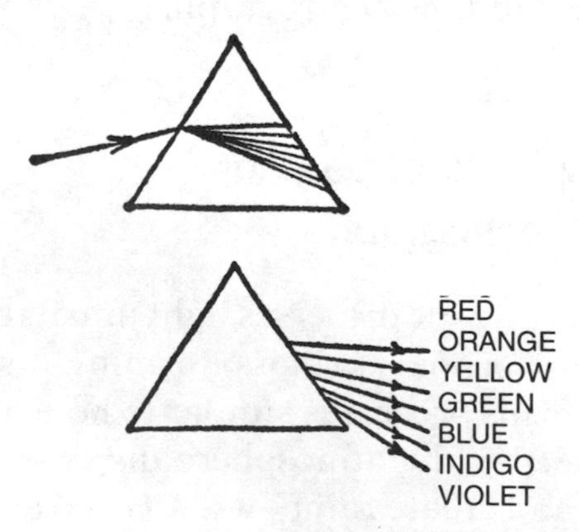

Fig. 3.16. The prism. Sunlight entering the prism is dispersed into a complete spread of colors called a spectrum. Leaving the prism, the colors are spread out still further. A screen held perpendicular to these rays would show a color parade similar to a rainbow.

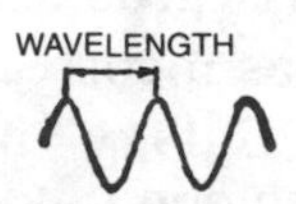

Fig. 3.17. Wavelength. Light may be thought of as waves. Technically light is known as one branch of a very large group called electromagnetic waves. As far as we know, the only basic difference between violet and blue light is in wavelength.

is stated in extremely small units called **angstroms.** One angstrom equals .00000001 of a centimeter and is abbreviated by the capital letter "Å". As stated previously, wavelengths of red light measure approximately 7,000 Å; the wavelength of violet light is about 4,000 Å; and, the human eye is most sensitive to wavelengths that measure around 6,000 Å (yellow-green).

Because they have different wavelengths, each color undergoes a different amount of bending (refraction) upon entering and leaving the glass prism. The short wavelengths (violet) are refracted more than the longer wavelengths (red). The several colors originally contained in the ray of white light are thus refracted by differing amounts in a process known as **dispersion.** This effect causes the chromatic aberration in refracting telescopes discussed earlier in this chapter.

Figure 3.18 shows the parts of the prism spectroscope:

1. A light baffle with a narrow slit through which the light enters. The narrowness of the slit prevents overlapping of the colors in the spectrum.
2. A collimating lens that arranges the light rays into parallel beams before they enter the prism.
3. The glass prism that disperses the light rays.

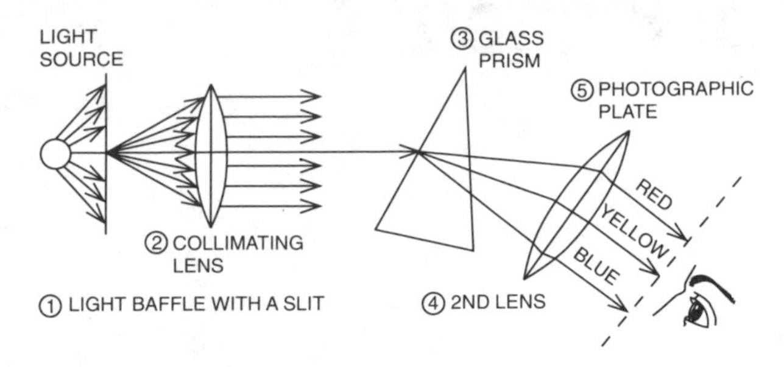

Fig. 3.18. The components of a complete spectroscope.

4. A second lens that focuses the dispersed light rays onto a photographic plate. There is one image formed on the plate, known as a spectral line, for each wavelength contained in the original light.
5. A photographic plate or electronic imaging device to make a permanent record of the spectrum.

Each element, such as oxygen, hydrogen, carbon, calcium, and so forth, whether found on the Earth or in deep space, leaves a telltale fingerprint on a spectrum. For example, oxygen, when heated in the laboratory, will emit the same spectral line as oxygen being heated on the surface of a remote star. By comparing spectral lines of an element in the laboratory with spectral lines obtained through a telescope's spectroscope, astronomers can identify the presence or absence of any and all elements in nature.

The Grating Spectroscope

A glass prism is not the only configuration of a spectroscope. More versatile is a device known as the grating spectroscope. Instead of a glass prism, a diffraction grating is used to disperse the light being analyzed. In its simplest form, a grating is a piece of glass on which a large number of parallel lines have been etched. The more lines the better, and good gratings may have as many as 40,000 lines per inch!

Astronomers favor the diffraction grating because it can create a much broader spectrum and separate very closely spaced spectral lines. This dispersion power helps the astronomer to identify the chemical composition of the star or galaxy under observation. In addition, gratings are sensitive to wavelengths in the infrared and ultraviolet range; whereas the prism works only in visible light.

The Message of Starlight

Between 1912 and 1914, the astronomer V.M. Slipher (1875–1969) studied the spectra of 15 nearby galaxies and found that the familiar spectral lines for hydrogen were all shifted toward the red end of the spectrum. His discovery showed that these galaxies were all moving away from us at tremendous speeds. In the 1920s and 1930s, astronomers Edwin Hubble (1889–1953) and Milton Humason succeeded in measuring the red-shift of many more galaxies. They discovered that, without exception, all the distant galaxies were receding from us at great speeds. The idea that the entire universe seemed to be exploding led to the theory of the expanding universe (Chapter 13).

CHAPTER FOUR

The Terrestrial Planets

KEY TERMS FOR THIS CHAPTER

aphelion
conjunction
elongation
greenhouse effect
inferior
inner planets
morning/evening star
opaque
opposition
outer planets
perihelion
superior
terrestrial
transit

The nine major planets of the Solar System are often divided into various groups. They can be classified according to their position to the Sun and the Earth. The **inferior** planets are Mercury and Venus, both inside the orbit of the Earth; all the other planets outside the orbit of the Earth are called **superior** planets. Another group divides the planets into **terrestrial,** or Earth-like planets, and the gas giants, or Jovian (Jupiter-like) planets. Or they can be grouped as **inner planets,** or those inside the Asteroid Belt (see Chapter 6), and **outer planets,** or those outside the Asteroid Belt (Figure 4.1). This chapter will examine the three inner planets—Mercury, Venus, and Mars. More complete numerical data for all the planets are found in Appendix A.

Mercury

Planetary Observations

Mercury is the planet nearest to the Sun. It can be observed in the sky with the naked eye only when the bright Sun is well below the horizon. The true orbit of Mercury follows an ellipse. An observer on the Earth sees this ellipse almost edgewise, making Mercury appear to move back and

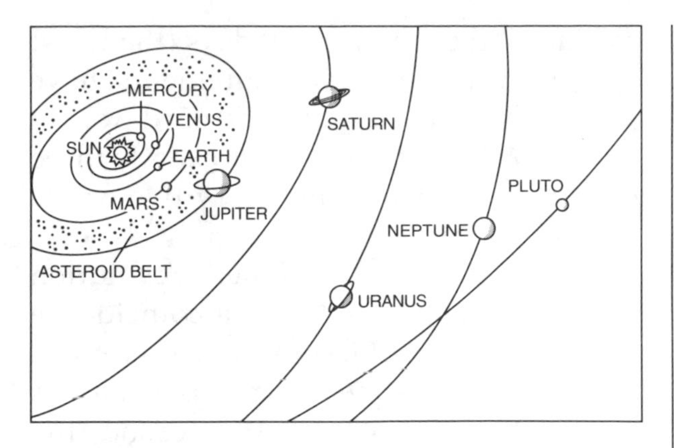

Fig. 4.1. Relative positions of the planets (not drawn to scale).

forth (east and west) of the Sun along a straight line. When the planet is west of the Sun, it is seen just before sunrise; when it is east of the Sun, it is seen just after sunset.

The synodic period of Mercury is 116 days. Half of this time Mercury leads the Sun, or it is west of the Sun; the other half, the planet trails the Sun. The ancient Greeks believed that they were observing two different planets. They used the name Mercury, the messenger of the gods, for the object they saw after sunset and they gave the name Apollo, the god of the Sun, to the object they saw before sunrise.

The farthest that Mercury can be on either side of the Sun is 28°. This angular distance is called **elongation,** or the angle (as viewed from the Earth) described by the Sun and one of the planets. The elongation of Mercury is never perpendicular to the horizon but somewhat tilted toward it. The planet is best seen as an **evening star** during March and April or as a **morning star** during September and October. These are the times when Mercury's elongation is closest to being vertical. Naked-eye observations of Mercury, under favorable conditions, reveal the planet as a medium-bright point of light. Twinkling, due primarily to the smallness of Mercury's disk and to its closeness to the horizon, can also be observed.

Through a small telescope, Mercury appears as a diffuse, whitish object going through phases. With a larger telescope a number of gray markings are visible on the planet's surface. Astronomers do not limit their observations of Mercury to the twilight hours. They use special filters to eliminate the diffuse sunlight from the telescope and are thus able to make observations of the planet in the daytime, particularly when Mercury is far to one side of the Sun. Greater accuracy in observation is possible when the planet is high in the sky instead of near the horizon, as its light is less affected by the Earth's atmosphere.

Phases of Mercury

Mercury goes through a series of phases, from full to new, similar to those of our Moon. When it is beyond the Sun, most or all of its illuminated face can be seen with a telescope but when it is on the same side of the Sun as the Earth, only a small crescent is visible. The brightness of the planet varies with the changing phases: It is brightest at crescent phase when it is closest to the Earth and less bright at full phase when it is closest to the Sun.

Transits of Mercury

Occasionally, Mercury passes between the Earth and the Sun. Such a crossing is called a **transit;** but it is not like a solar eclipse, as the planet covers only a very small portion of the solar disk. Transits can only be observed with the aid of a telescope. The observer sees a small, black circle, less than 1 percent the diameter of the Sun, slowly crossing the solar surface.

Transits only occur either in May or November. (Another transit of Mercury will take place on November 14, 1999.) Transits occur rarely because of the large (7°) tilt of Mercury's orbit to the orbit of the Earth, causing the planet to pass to the north or south of the Sun. Accurate measurements of transits are used not only in exact determination of the orbit of Mercury, but also in computing the Earth's period of rotation. Careful analysis has shown that the Earth's rotation is actually slowing down. The period of one rotation about Earth's axis will increase by one second in the next 100,000 years.

Rotation and Revolution

Initially, astronomers had difficulty determining Mercury's actual period of rotation mainly because of its small size. The best estimate available was a period of 88 days—an estimate resulting from G.V. Schiaparelli's (1835–1910) observations in 1889. This period of 88 days is also equal to Mercury's sidereal period of revolution around the Sun. Under these conditions, one side of Mercury always faces the Sun, as one side of our moon always faces the Earth. In 1965, astronomers G.H. Pettergill and R.B. Dyce bounced radar signals off Mercury using the 1,000-foot radio telescope in Arecibo, Puerto Rico, and discovered that the planet rotates about its axis around once every 59 days (the latest and more accurate measurements have shown that the period is 58.65 days). In other words, Mercury makes exactly three rotations on its axis for every two orbits around the Sun.

This 3:2 ratio of the planet's revolution to rotation is probably not a coincidence for two reasons: First, Mercury is not a perfect sphere—one of its diameters is longer than the others. And second, the gravitational pull of the Sun, which slowed the planet's spin, keeps its elongated diameter at **perihelion,** or the point in a planet's orbit furthest from the Sun, and in line with the Sun for every 1.5, 3, 4.5, 6, and so forth rotations of the planet (Figure 4.2).

A complete sunrise-to-sunrise day on Mercury would last two of its years, or 176 Earth days. Another effect is that, while closest to the Sun, the increased orbital speed of the planet overtakes the Sun's apparent motion in the sky. During perihelion passage, an observer on Mercury would observe (depending on his or her

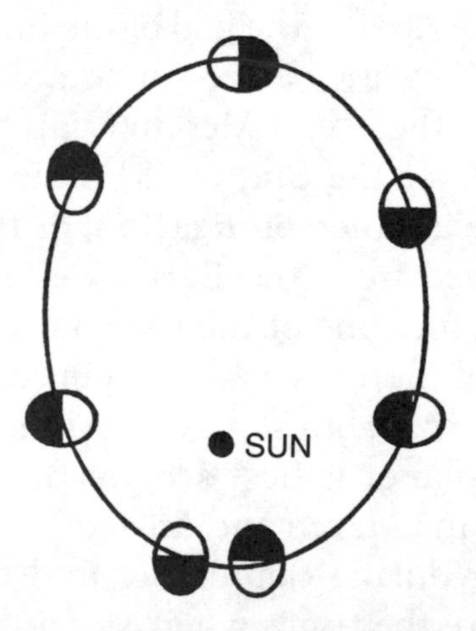

Fig. 4.2. The revolution (around the Sun) and the rotation of Mercury. Note that the bulge of the elongated diameter of Mercury is again in line with the Sun after 1.5 rotations.

location on the planet's surface) a double sunrise, a double sunset, or see the Sun backtrack in the sky at noon.

Temperature and Albedo

Mercury is a planet of extreme temperatures because of its lack of atmosphere and its closeness to the Sun. At noon, at the planet's equator, the temperature on the surface reaches close to 800° F—a temperature at which lead and zinc would flow like water. Just before Mercury's dawn, the temperature hovers around −300° F. Radio measurements taken just below the surface indicate that at about three feet, the temperature is at a constant 170° F.

Mercury's albedo, or ability to reflect light, is .06. This means that 6 percent of the sunlight received by Mercury is reflected back into space, and that 94 percent of that light is absorbed by the planet's surface.

Surface Gravity and Escape Velocity

The surface gravity of Mercury is about .37 times, or around one-third, that of the Earth. A 90-pound person would weigh 30 pounds on a scale on Mercury. Because of the low surface gravity, the escape velocity from the surface of the planet is only 2.6 miles per second. Any object leaving Mercury with a velocity of 2.6 miles per second would leave forever. To compare, on Earth, the escape velocity is 7 miles per second.

Gases tend to be expansive—that is, every molecule of gas spreads in all directions and occupies larger volumes. The surface gravity of the Earth is strong enough to counteract the atmospheric gases' expansiveness. On Mercury, however, the gravitational force is too weak to prevent the escape of gas molecules with speeds in excess of 2.6 miles per second. Because of this, if there had been an atmosphere on Mercury millions of years ago, it must have escaped. Yet astronomers believe that there is an extremely thin atmosphere surrounding Mercury. From data obtained by the Mariner 10 mission in 1974, it was learned that the solar wind acting upon Mercury's magnetic field produces an almost vacuum-like atmosphere made up of hydrogen, helium, and oxygen. In 1985 through 1986, astronomers at the McDonald Observatory also discovered the presence of sodium and potassium in the atmosphere. However, the whole atmosphere is so thin that its atoms rarely collide with one another, and the atmospheric pressure is only one-trillionth that of the Earth's at sea level!

Missions to Mercury

In March 1974, the United States successfully completed a flyby of Mercury with the robot spacecraft, Mariner 10. From a distance of less than 500 miles, this spacecraft photographed one hemisphere of the planet, taking images of around 50 percent of the planet.

Much of what we know about Mercury is from this mission. At first glance, the surface of Mercury greatly resembles that of the Moon. There are many heavily cratered upland regions and large smooth plains that surround and fill impact basins. Unlike the Moon, however, there are large regions of gently rolling plains, a global system of thrust faults formed by the shrinking of the planet as the interior cooled, and cliffs hundreds of miles in length and several miles high. The largest structure viewed by the spacecraft was the Caloris Basin, an enormous impact crater 807 miles in diameter (the size of Texas), surrounded by mountain blocks reaching up to 1.2 miles above the surrounding plains. Mercury is often called "the metallic planet" because Mariner 10 also discovered that Mercury has a large iron core—the largest in volume compared to any other planet in the solar system. Astronomers are not yet certain why this is so, but believe that finding the answer may unlock the mystery of Mercury's origin and the origin of the other terrestrial planets as well.

Venus

Planetary Observations

Venus, named after the Roman goddess of love and beauty, is the second planet from the Sun. It is also the brightest of the heavenly objects, with the exception of the Sun, the Moon, and extremely bright, rare comets. At its brightest, it is visible during the daytime (when not too close to the Sun) and is strong enough to cast shadows after dark. The synodic period of Venus is 584 days. Half of this time (a bit less than 10 months) it rises ahead of the Sun; and the other half, it sets after the Sun. As in the case of Mercury, the ancients thought that they were observing two different objects. They referred to the morning star as Lucifer and to the evening star as Hesperus.

The orbit of Venus is an almost perfect circle, with an eccentricity of only .007. Therefore, as seen from the Earth, Venus seems to travel back and forth on an almost straight line east and west of the Sun. Venus' greatest elongation from the Sun is about 46°, allowing it to set (or rise) almost 3 hours after (or before) the Sun. Morning observations in the fall are the most favorable in the northern hemisphere when elongation is the greatest and the glare of the Sun does not overwhelm Venus.

To the unaided eye, Venus appears as a bright object shining with a steady, white light. At its brightest, Venus reaches an apparent magnitude of −4.4. To compare, the magnitude of the Moon is −27; Venus is also 12 times brighter than Sirius, the brightest star in the northern hemisphere.

Because of its dense cloud cover, Venus reveals very little information—even with the more powerful telescopes. In fact, until robot spacecraft visited the planet, astronomers could only see the very tops of its thick cloud cover. Venus remained a mysterious planet—bright, diffuse, and without any permanent markings. Scientists only knew that Venus underwent a complete series of phases like Mercury and the Moon.

Phases and Brightness

As Venus approaches a line joining the Earth and the Sun, it reveals a crescent, which grows narrower as it approaches—the length increasing as the width of the planet decreases. Just before the light is entirely cut off from Venus (as with a new Moon), its crescent, as seen from Earth, is six times larger than when it is full. When the planet is directly between the Sun and the Earth, it is said to be at inferior conjunction. Venus is full (as with a full Moon) when it is on the opposite side of the Sun from the Earth, or at superior conjunction (Figure 4.3).

Venus' remarkable brightness results from two favorable factors. It is close to the Sun and to the terrestrial observer. It has a high albedo value, or reflectivity. Nearly 80 percent of the light received by Venus from the Sun is reflected back into space.

Venus does not appear at maximum brightness in its full phase because of its greater distance from the Earth at that time. It appears brightest just before and just after it passes inferior conjunction. At this point, it is close to the Earth with the crescent wide enough to supply a sizable reflecting surface.

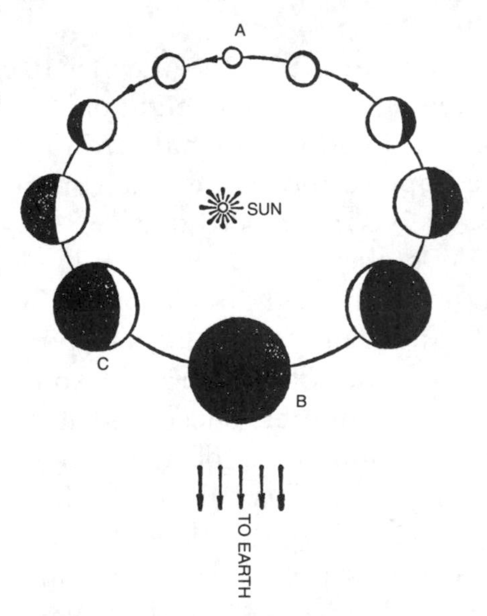

Fig. 4.3. Phases of Venus. Venus appears as a circular disk of light (full) when it is at superior conjunction (A on diagram).

As it moves toward position B (inferior conjunction), less and less of the illuminated surface of Venus can be seen by a terrestrial observer.

However, because its diameter increases, Venus will appear at its brightest 36 days before (position C) or 36 days after inferior conjunction.

Transits

On rare occasions, Venus passes directly in front of the Sun. Such a transit may be observed only through an approved solar filter. NEVER LOOK AT THE SUN WITHOUT SUCH A FILTER OR PERMANENT BLINDNESS WILL RESULT! As with transits of Mercury, there is no solar eclipse because Venus' shadow covers only a small portion of the Sun's face. If the transit is centrally located, the crossing may last as long as 8 hours. The first such transit of Venus was first accurately described in 1639. Future transits will occur in 2004 and 2012.

Transits of Venus do not happen at every inferior conjunction because of the 3° inclination of the planet's orbit to the ecliptic. As a result, Venus is either on one side or the other side of the Earth's orbit. In other words, half of the time it is slightly above or below the ecliptic. It is only when Venus is just crossing the ecliptic and at inferior conjunction that a transit is possible. The two conditions happen together an average of once every 50 years.

Rotation and Revolution

Because there are no fixed markings in the clouds surrounding Venus, and the permanent cloud cover makes it impossible to see the surface, it took a long time to determine the rotational periods of the clouds and the planet. In the 1950s, using improved radar techniques, it was found that Venus rotates in a clockwise or retrograde direction once every 243 days—opposite of most of the other planets and satellites in our solar system. Also knowing that Venus takes 584 days to complete one revolution around the Sun, astronomers determined that Venus presents the same face to the Earth at every inferior conjunction. It was also found that Venus' axis is only slightly tilted (3°) to its orbital plane, as compared to the Earth's axis tilted at 23½°.

Space-age Observations of Venus

Until fairly recently, detailed information about Venus was not available or even certain. However, knowledge of the planet improved in the early 1960s. The following is a brief chronology of the missions dedicated to the enhancement of our knowledge of Earth's sister planet and the data they obtained:

• In 1961, the first successful radar observations of Venus were made by a joint team from the United States, the Soviet Union, and Great Britain at Jodrell Bank in Great Britain.

• In 1962, astronomers from the Jet Propulsion Laboratory in California, using the giant antenna at Arecibo, Puerto Rico, bounced radar signals off the surface of Venus. By analyzing the returning radio signals, they discovered two bright areas on the surface. They named them Alpha and Beta ("bright" to radar means higher elevation).

• In December 1962, the U.S. space probe Mariner 2 conducted the first successful flyby of Venus, passing within 22,000 miles of the planet. It confirmed that the planet is completely covered with yellowish-white clouds, with no glimpse of any solid surface through a break in the clouds.

• The Soviet Union launched a series of probes to Venus they called Venera. Venera 4 was launched in June 1967; Venera 5 in January 1969; and, Venera 6 also in January 1969. Although these missions were short-lived, they were successful in sending back useful information.

• In 1972, the Soviet probe, Venera 7 was the first to soft land on Venus and revealed a surface temperature close to 900° F, with only a small difference in temperature between the sunlit and dark sides of the planet. It also measured an atmospheric pressure 90 times that of the Earth's. In 1972, Venera 8 was the second successful, though short-lived, landing on Venus, and found that the soil was like granite.

• In February 1974, the Mariner 10 craft investigated Venus at a height of less than 4,000 miles and found that strong winds in the upper atmosphere reach speeds in excess of 200 miles per hour.

• On October 22, 1975, a probe launched from the unmanned Soviet spacecraft Venera 9 landed on Venus and, surviving the lethal extremes of temperature and pressure for 53 minutes, succeeded in transmitting photographs of the immediate landscape, as well as a flood of other data from the surface of the planet. Three days later, a similar probe from the twin craft, Venera 10, landed 1,375 miles away from the Venera 9 probe and returned its findings for 65 minutes. These were the first photographs taken on the surface of another planet.

Preliminary studies of these data revealed a great deal of information. The probes found an extreme value for atmospheric pressure—90 to 92 times the Earth's surface atmospheric pressure; the surface temperature was reported to be 905° F by Venera 9 and 870° F by Venera 10; the surface wind speeds were found to be very low, ranging at both sites from 2 to 6 miles per hour; the amount of sunlight transmitted through the clouds turned out to be greater than originally expected, with shadows of objects on the surface clear and sharp on the photographs; and jumbles of large rocks (12 to 16 inches across) were strewn over the landscape—an area previously thought to be a sandy desert. The rocks at the Venera 9 site included both smooth, rounded specimens as well as angular rock fragments, while the ones near Venera 10 were flat pancake-type stones.

- In 1978, NASA launched two Pioneer Venus Orbiters. Pioneer Venus 1 completed the first global radar mapping of the surface, successfully surveying 90 percent of the planet's surface in less than two years. Two giant upland areas were found, Ishtar Terra and Aphrodite Terra, as were great mountains and rift valleys. Pioneer Venus 2 sent four probes toward Venus, sending back data for 68 minutes as it descended through the thick atmosphere.
- Four more Venera missions followed in 1982, 1983, and 1984. Venera 13 and 14 launched two more probes that made crude chemical analyses of the Venusian soil. Venera 13 also took the first colored photograph of the Venusian surface. Venera 15 and 16 orbited the planet and used a new type of radar called synthetic aperture radar (SAR) to bring out finer detail in images. Two more landers reached Venus in 1985, Vega 1 and 2. By that time, the Soviets had gathered the most complete data on Venus to date.

The Magellan Mission

The most ambitious and sophisticated survey of the surface of Venus began September 1990 when the radar mapping satellite Magellan (originally called the Venus Radar Mapper) turned on its own version of synthetic aperture radar. Magellan's radar resolves surface features as small as 360 feet across—about 10 times better than the images resolved by the Venera missions.

Here are some of the data obtained from Magellan:

- Venus has no craters smaller than about 4 miles in diameter. This is because the atmosphere is 90 times denser than the Earth's, stopping all incoming objects that would make smaller craters.
- Venus' rocks deform easily because they are heated halfway to their melting points by the 900° F temperature at the surface.
- There has been, and still may be, widespread volcanic activity on Venus that has produced vast lava plains and channels similar to those on the Moon. Other volcanic features are unique, including arachnoids (named because of their spider-like appearance) formed when molten material upwelled near the planet's surface. Numerous domes, some that look like huge pancakes, were also formed by volcanic activity.
- There are large areas of fractured terrain that indicate tremendous stresses in the planet's crust. Moving molten material and crustal movements have caused a great deal of faulting in Venus' morphological history.

• There seems to be very little erosion taking place on Venus' surface. The surface may be very young—with estimates ranging from 400 million to 1 billion years old.

• Most ejecta, the blanket of material thrown out from an impact, is asymmetrical around Venus' craters. The thick atmosphere prevents the outward-flying debris from creating a circular blanket, such as seen on the Moon, around the craters.

The Magellan spacecraft has already mapped 100 percent of Venus' surface—much of which has never been probed before. If Magellan remains mechanically healthy, it has enough fuel to allow the mapping mission to continue for decades to come.

Today's Venus

Not too long ago, there was a common belief (maybe more of a wish) that Venus was truly Earth's twin. This idea pictured Venus as a more primitive Earth, with vast jungles, large bodies of water, and perhaps, some form of early life—all under an oxygen-rich, moist atmosphere.

Today, however, Venus is known to be a planet with hellish conditions—a crushing, opaque atmosphere of carbon dioxide clouds, sulphuric acid rain, violent lightening, temperatures that average 900° F, and a landscape that is a barren and rocky furnace.

Why is Venus so hot? Scientists believe it is due to the **greenhouse effect.** The greenhouse effect is best demonstrated in Earth's greenhouses. The glass of the roof and sidewalls act as an energy trap, efficiently trapping the Sun's radiant energy—mainly at wavelengths between 4,000 and 7,000 angstroms—to which the glass is perfectly transparent. The glass does not allow the incoming radiant energy of the warmed air to escape, because it is **opaque** to that wavelength (around 100,000 angstroms). This causes energy, in the form of heat, to build up in the greenhouse. This effect is clearly felt when you enter a closed car that has been standing in the heat on a summer's day.

In the case of Venus, the great amount of carbon dioxide (CO_2) in the atmosphere acts like the glass of a greenhouse, thus preventing this radiant heat from escaping back into space. Because of the atmosphere's circulation, the incoming solar radiation is able to spread through the clouds and onto the surface. However, the Sun's radiation grows weaker as the density of the atmosphere increases. At the surface, the intensity of the sunlight is equal to a heavily overcast day on the Earth. Therefore, surface temperatures are somewhat lower than in the upper atmosphere.

Mars

Mars is the fourth planet from the Sun, and therefore, its orbit lies outside that of the Earth's. Once every synodic period (about 780 days), Mars and the Earth are in line with the Sun and on the same side of the Sun; or, they are in line on opposite sides of the Sun. This is primarily due to the longer distance that Mars must travel to complete one orbit around the Sun.

The first event is called **opposition,** when Mars and the Sun are on opposite sides of the Earth. Mars and the Earth do not follow equidistant tracks because of variations in the Martian orbit. Their distances at opposition between them varies from a minimum of 35 million miles to a maximum of 63 million miles (Figure 4.4). The two planets are closest around August 25 and most distant in February. Oppositions on or near the August date are called "favorable oppositions" and occur at intervals of 15 and 17 years. Before the use of space probes, much of our knowledge about Mars was obtained during such favorable oppositions. In July 1986 and September 1988, two such oppositions took place. The next one will occur in 2003.

The second event is called **conjunction,** when the Sun and Mars are in the same direction as viewed from the Earth. At conjunction, Mars is lost in the Sun's glare and is not visible. The farthest that Mars can be from the Earth is at conjunction, when the distance is 247 million miles—more than seven times farther than at favorable opposition.

Mars has long been an object of intense fascination for both the astronomer and the general public. It is the only planet that, until very recently, held the possibility of life. In 1894, an American astronomer, Percival Lowell (1855–1916), built the Lowell Observatory in Flagstaff, Arizona, specifically to search for the presence of Martian life. Detailed drawings of his observations seemed to indicate that an intelligence had constructed a complex system of canals, from the polar cap to the equator, for the purpose of irrigation. His findings created a heated debate in the astronomical world, with other noted astronomers denying that any such structures existed. Popular writers of the time (notably H.G. Wells, who wrote *The War of the Worlds*) joined the fray and created works of fiction portraying Martian beings. Modern films and television have continued the tradition to this day—even though we now know there are no Martians on Mars.

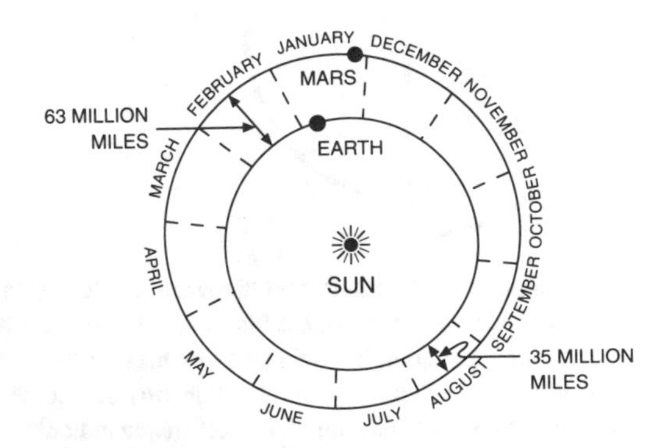

Fig. 4.4. The orbits of Mars and Earth.

Planetary Observations

To the naked eye, Mars looks like a point of light with a distinct reddish-orange color. Its motion follows the ecliptic very closely, as the planet is tilted only 1.9° to the plane of the ecliptic. Mars varies in brightness as it approaches or recedes from the Earth. When closest to the Earth (35 million miles), Mars shines more than 50 times brighter than when it is farthest away (247 million miles).

The best telescopes show that the surface of Mars is a patchwork of several colors, including red, gray, and white. The reddish areas are presumably deserts; their color is uniform, and they extend over 60 percent of the planet's surface. The gray areas, covering about 40 percent of the planet's surface, changes from gray in the Martian winter to blue-gray in the summer. The reddish areas are more common in the northern hemisphere; while the gray is more prevalent in the southern hemisphere.

White patches are found near both the north and south poles of the planet and change with the local seasons. A white cap appears at the beginning of winter when a particular pole is pointing away from the Sun; it often disappears during the summer when the pole points toward the Sun. The north and south polar caps thus alternate in size depending on the season.

Each polar cap, composed of carbon dioxide (CO_2) ice, builds very rapidly. In several days it may extend some 20° to 30° from the pole. This period is followed by a long stagnation (the winter season), when it remains essentially unchanged. With the coming of spring, the cap contracts and breaks up into several small, irregular white spots. These spots disappear altogether with the approach of summer. In general, the south polar cap generally grows larger than the north polar cap; it is also more likely to disappear entirely during the southern hemisphere's summer.

The Atmosphere of Mars

Theories, experiments, and observations show that Mars has a definite atmosphere. Calculations have determined that the escape velocity for Mars is 3.2 miles per hour (remember Earth's is 7 miles per hour)—high enough for the red planet to have an atmosphere. Experiment and direct observation have also provided the following results:

- Sunlight on Mars is reflected and diffused similar to twilight on Earth.
- Mars is a better reflector of sunlight than an airless planet such as Mercury. The red planet has an albedo of .15—or 15 percent of the sunlight falling on Mars is immediately reflected back out into space.
- Photographs of the planet taken through ultraviolet filters show it to be larger than photographs taken through an infrared filter. The infrared filter records the red and infrared light coming directly from the surface of the planet; while the ultraviolet filter records the blue and ultraviolet rays scattered by the atmosphere. The difference between the two radii, around 60 miles, is assumed to be the thickness of the Martian atmosphere (Figure 4.5).
- The variation in the size of the polar ice caps, made mostly of frozen carbon dioxide (CO_2), is further direct proof that an atmosphere exists. If there were no atmosphere, the polar cycles of forming and melting would not be possible.
- Telescopic observations have revealed the presence of clouds and mists. Most convincingly, huge dust storms sometimes

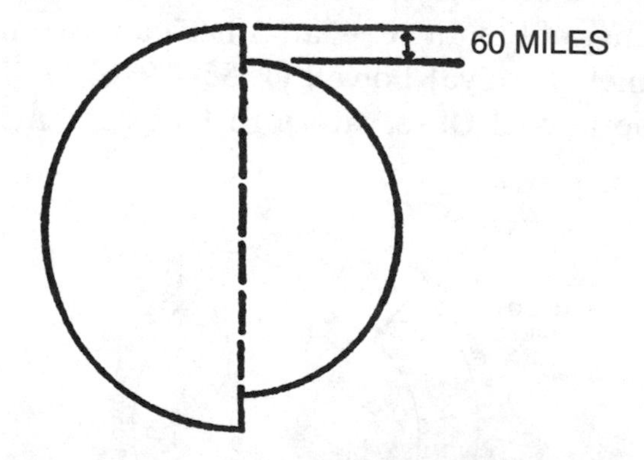

Fig. 4.5. Comparison of infrared and ultraviolet photographs of Mars. The left side was taken with a filter that lets through blue, violet, and ultraviolet light. The right side was taken with the aid of a filter through which red and infrared light passes. The difference in size is to be carefully noted. This difference indicates the thickness of the atmosphere that diffuses the blue and violet light.

cover most of the visible face of Mars, with some of the storms creating tall columns of dust called "dust devils."

Seasons and Climates

The seasons on Mars are in some respects similar to those on the Earth. The widths of the climactic zones—due primarily to the tilt of the planet's axis—are similar on both planets. (Mars is tilted at an angle of 25.2°, as compared to 23.5° for the Earth.) In addition, both planets have four seasons. The variations in the periods of daylight are also similar on both planets, with the Martian day slightly more than 24½ hours as compared to our 24-hour day.

There are also several differences between the Earth and Mars. Each season on Mars is twice as long as each season on the Earth, especially since the length of the season depends on the period of one revolution around the Sun. Earth's sidereal period or calender year is 365¼ days, while it is 687 days for Mars. The average temperatures in each of the climate zones on Mars are less than those of the corresponding zones on Earth. This is because Mars is about 50 percent farther away from the Sun than the Earth, so the intensity of the Sun is much less on Mars. In addition, the thin Martian atmosphere does not hold the heat very well. The summer temperature in the equatorial zone of Mars at noon seldom rises above 32° F; while nighttime temperatures plunge to more than −130° F. In addition, summer in Mars' southern hemisphere is much warmer than summer in its northern hemisphere because of the large variation in the planet's orbit. Mars is closer to the Sun by 20 percent at its perihelion than at its **aphelion,** or the point in its orbit closest to the Sun. The difference for the Earth is only 3 percent.

Space-age Observations of Mars

What were these changing colors and features on Mars? In order to obtain a good look at the planet, scientists began to send numerous spacecraft. The following is a brief chronology of these missions:

- Most of our present-day knowledge about Mars began with a series of Mariner missions from 1965 to 1969. Mariner 4 flew by the red planet in July 1965, took 22 photographs and discovered a crater density similar to that of the lunar highlands. In 1969, Mariners 6 and 7 found similar terrain during their flybys. Besides looking more and more like the cratered Moon, the spacecraft found a polar cap of dry ice, a giant crater called Nix Olympia, and haze, clouds, and wind. However, these missions only covered the southern hemisphere, which, during later missions, was found to be very different from the northern hemisphere.

These missions also revealed that the Martian atmosphere is very thin—about $1/200$ the density of the Earth's atmosphere (or equal to the pressure of the Earth's atmosphere at an altitude of 100,000 feet). A major part of the Martian atmosphere was found to be carbon dioxide (CO_2) with traces of water vapor and oxygen.

- In 1973, the findings of a short-lived Russian surface probe, the Soviet 5 Mars Probe, suggested that as much as 30 percent of the Martian atmosphere may be the inert gas argon.

• In November 1971, the United States successfully placed Mariner 9 in orbit around Mars. But the spacecraft was greeted by a major dust storm that covered the whole planet. The dust storm lasted for nearly eight weeks, then slowly disappeared. After the dust settled, Mariner 9 continued with its work. Among its many discoveries was the presence of huge volcanoes. One is called Olympus Mons, the largest volcano in the solar system, towering almost three times as high as Mt. Everest and its base covering an area equal to the distance between San Diego and San Francisco. The spacecraft also sent back images showing detailed features, such as long channels resembling terrestrial riverbeds (dry) and flood plains, and eroded craters. The surface of Mars also revealed many canyons, one of which, Valles Marineris, is more than 2,000 miles long, 75 miles wide, and 4 miles deep. There were featureless deserts and dune-like areas greatly resembling sand dunes on the Earth. Mariner 9 also found that the Martian dust clouds are often stationary for long periods of time. They have been known, though, to move across the Martian deserts at speeds in excess of 200 miles per hour. The data seems to indicate that dust storms often occur when Mars is closest to the Sun.

The long river channels have led scientists to speculate that the northern hemisphere may have had running water in the past. Some astronomers have even suggested that a shallow sea once covered much of the northern hemisphere of Mars. The burning question for now is, What happened to all that water?

The findings of the Mariner missions indicate that the surface of Mars is both lunar-like and Earth-like; both dead and alive. Like the Moon, many craters dot the Martian landscape. Some of the smaller craters form lines and could have easily been mistaken for artificial channels by early observers. There are also many hills that wind across the surface like the rills on the Moon. The Martian craters, however, show definite signs of having been eroded by the Martian weather, much like erosion on Earth; lunar craters do not.

Other data obtained from the Mariner missions showed that Mars has a very weak magnetic field (less than 1/1000 of the Earth's), suggesting that the core of the planet is solid. In addition, the Mariners found no evidence of a radiation belt around Mars.

• Two U.S. scientific vehicles called Viking 1 and 2 were designed to search for life on Mars. Viking 1 touched down on the Martian surface on the morning of July 20, 1976; Viking 2 followed on September 3, 1976. As astronomers held their breath, the first pictures of the Martian landscape were sent back to Earth from the Viking 1 lander—a rocky field stretching to the horizon.

After landing, the robot arms on both crafts scooped up samples of the Martian soil and analyzed it for signs of present or past life. Some of the findings escape full explanation to this day, but in the end the Vikings did not find any definite evidence of life—at least not in their little corner of Mars.

The Moons of Mars

Mars has two tiny satellites—Phobos ("fear") and Deimos ("panic"), named for the two mythological companions of Mars, the god of war. Both are tiny moons—often described as potato-shaped. Phobos, the larger, has an average diameter of about 10 miles, while the average diameter of Deimos is only 7 miles. Both revolve around the equatorial plane of Mars in the usual counterclockwise direction.

Phobos is 5,800 miles from the center of its parent planet and a mere 3,700 miles above its surface. It revolves around Mars in 7 hours, 39 minutes—less than the Martian day! Phobos rises in the west and sets in the east 4½ hours later. Deimos is 14,600 miles from the center of Mars. It revolves around the planet in 30 hours, 18 minutes (six hours longer than the Martian day) and, therefore, goes through two complete cycles of phases before it sets in the west.

Astronomers believe that these two moons may actually be captured asteroids. They calculate that probably within a few million years, the gravity and atmosphere of Mars will eventually bring these two satellites crashing to the Martian surface.

CHAPTER FIVE

The Outer Planets

KEY TERMS FOR THIS CHAPTER

Jovian	*planetesimals*
oblate	*retrograde*
occult	*solar nebula*
organic	*zones*

The outer planets are the gas giants, or **Jovian** planets of Jupiter, Saturn, Uranus, and Neptune. There is one exception to the outer planets: Pluto, a small, dense, rocky, solid planet, much like the inner, terrestrial planets. The gas giants are huge and primarily made of gas with little or no solid surfaces. These planets formed under very different conditions in the outer **solar nebula** (gas and dust from which the Sun and planets formed) 4.6 billion years ago. Each of these giant planets has a complex system of rings and moons that may hold the key to understanding of how our solar system developed over time.

Much of our current data on these planets was obtained by two very successful Voyager missions. Voyager 1 and 2 were launched in 1977 and spent the next 12 years on a grand tour of the gas giants, sending back to Earth astonishingly detailed images. Just analyzing the thousands of images will keep planetary scientists busy for decades.

Jupiter

Planetary Observations

To the unaided eye, Jupiter, named for the chief of the Roman gods, appears as a bright yellowish object. It travels slowly through the 12 constellations of the zodiac and completes the circuit in just under 12 years. Only one star, Sirius, and the planet Venus outshine Jupiter (the planet Mars, at some favorable oppositions, also is brighter). For nearly six months of every year, Jupiter is visible every night—a brilliant planet in a field of stars.

Through even a modest-sized telescope, Jupiter appears as an **oblate,** or pear-shaped, sphere, with prominent belts—either orange, brown, white, or red—running parallel to its equator. What the observer sees is really the top of the very thick Jovian cloud cover. These belts are thought to be caused by different underlying currents of the planet's very deep atmosphere—flowing alternately east and west in both hemispheres. They resemble the trade wind zones of the Earth. The equatorial belt is light in color and varies from 12,000 to 15,000 miles in width. On each side of the equatorial belt, alternate dark and light belts (the light belts are called **zones**) run parallel to each other. While maintaining their general outline over time, the belts and zones vary in location, color, and form. When viewed over the course of several days, the movements and changes of these belts and zones display tremendous energy in the Jovian atmosphere.

Rotation and Revolution

The rotational period of Jupiter varies with the latitude—it is shortest at the equator (9 hours, 50 minutes, 30 seconds) and longer north and south of the equator. Variations of rotational period do not increase uniformly in latitude but are rather uneven. To complete the confusion, the periods of rotation of like latitudes in the northern and southern hemispheres are not equal to one another; nor do the periods remain constant for any length of time.

Jupiter's rotation has been determined from observations of the semi-permanent markings on the belts and zones of the planet. In addition, the Great Red Spot (GRS), a huge hurricane-like storm, has been observed for nearly 300 years and has provided a very reliable marker.

Jupiter has a radius of 44,380 miles and a period of rotation just under 10 hours. This suggests that material at the top of the cloud deck (all that is visible) must whip around the axis at speeds near 30,000 miles per hour. In comparison, the Earth's surface rotates at a mere 1,000 miles per hour.

The high speed of Jupiter's rotation is no doubt responsible for its considerable equatorial bulge. Jupiter's oblate shape is easy to measure: Its diameter through the equator is nearly 6,000 miles longer than through the poles.

Jupiter is one of the superior planets and orbits beyond Mars and the asteroid belt (see Chapter 6) at a mean distance of 467 million miles or 5.2 AU from the Sun. The planet takes 11.86 years to make one revolution around the Sun at a mean velocity of 8.1 miles per second. Jupiter's axis is tilted a little over 3° to the plane of its orbit (the Earth's tilt is 23½°), and its orbit is tilted slightly more than 1° to the plane of the ecliptic.

The Structure of Jupiter

Jupiter is by far the largest planet in our Solar System, with a mean diameter of

85,680 miles (the mean diameter of the Earth is only 7,900 miles). In other words, it would take 11 Earths lined up side by side to equal Jupiter's diameter, and 1,330 Earths to fill the same volume. Jupiter occupies the same space as all the other planets combined with some room to spare. Its mass, too, is enormous: More than 300 Earths would be needed to balance the mass of Jupiter.

Jupiter may hold the key to understanding the nature of the origins of our planetary system because it still has many of its primordial material left over from the formation of the Solar System. Jupiter (as well as the three other gas-giants) exhibits certain characteristics very different from the terrestrial planets:

• Jupiter is much larger than any of the terrestrial planets (Mercury, Venus, Earth, Mars, and Pluto).

• Jupiter is less dense than the terrestrial planets, being only about ¼ the density of the Earth.

• Most of its volume is made of hydrogen (like the Sun) in both a liquid metallic and gaseous state. The liquid metal core of hydrogen is believed to extend out to a distance of 40,000 miles from Jupiter's center. These gases may surround a small rocky core (Figure 5.1).

• Theoretical estimates suggest that the pressure at the center of Jupiter is around 10 million pounds per square inch with temperatures as high as 50,000° F.

• A thick layer of compressed gaseous hydrogen (82 percent) and helium (17 percent) surrounds Jupiter's liquid metal core. The remaining 1 percent is primarily ammonia (NH_3), water (H_2O), and methane (CH_4). This envelope is topped by two

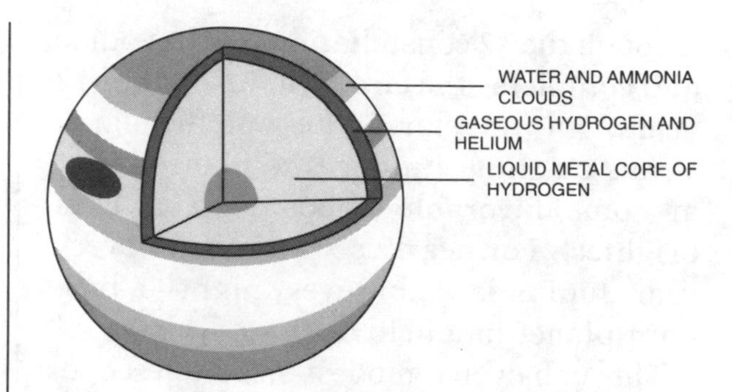

Fig. 5.1. The structure of Jupiter. Jupiter's dynamic atmosphere produces the distinct bands and zones visible to the observer.

thick cloud decks. The lower deck is made of water droplets and ice crystals.

• Jupiter's upper-cloud deck consists of crystals of frozen ammonia compounds. The temperature at the top of these clouds is around −300° F. Its colorful belts and zones are the visible face of the planet and are probably created by a combination of the planet's rapid rotational spin, the internal heat from the planet's interior, and the chemical make-up of its atmosphere. Scientists assume that the temperature rises below the level of the cloud tops, just as on any planet with a dense atmosphere. This is primarily due to the greenhouse effect.

• White ovals and spots, large brown ovals, white plumes, and giant lightening flashes have been observed for many years.

• In addition, there is the Great Red Spot (GRS)—the semi-permanent storm center or hurricane that has been observed for 300 years. Its brick color periodically fades and then returns; and, the GRS varies in size, reaching a maximum length of over 24,000 miles and a maximum width of over 7,800 miles—3 times the diameter of the Earth! The GRS, however, is unlike any terrestrial storms since it seems to remain fixed in place in relation to the surrounding cloud features, and it has persisted for so long; hurricanes on Earth tend to wander and last for only a few days. Jupiter's atmosphere is literally a seething cauldron of activity.

• Jupiter has a thin but definite system of rings and at least 16 satellites orbiting the planet like a miniature solar system.

The Formation of Jupiter

As the planets formed 4.6 billion years ago, cooler temperatures in the outer solar nebula prevented much of the primordial hydrogen and some helium from being heated and vaporized. The gravitational pull of this larger mass of gas was able to attract even more hydrogen and helium from neighboring space, contributing to the gaseous atmospheres of the outer planets—including Jupiter.

As gravity condensed Jupiter, the planet began to heat up and give off its own energy. Today, measurements show that Jupiter radiates about 1.7 times more energy than it receives from the Sun. One theory is that if Jupiter had been 50 to 85 times its present mass, it would have become a companion star to our Sun—with disastrous results for future life on Earth. (A more detailed examination of the formation of the Solar System is found in Chapter 8.)

The Satellites of Jupiter

Jupiter is known to have at least 16 moons. The four largest and most interesting satellites were discovered in 1610 by Galileo Galilei and are called the Galilean moons. In increasing distance from Jupiter, they are Io, Europa, Ganymede, and Callisto. These moons can be seen even through the most modest pair of binoculars or small backyard telescope.

Interestingly, the four Galilean satellites follow thc same pattern of density and size as the Solar System: That is, the satellites' densities decrease as their distance increases from Jupiter. In addition, the two outer moons, Ganymede and Callisto, are larger than the inner two moons, Io and Europa. Ganymede, Io, and Callisto are larger than our Moon and Ganymede is larger than the planet Mercury.

Early in its formation, Jupiter gave off enough energy, causing the planet's inner satellites to form dense, rocky, small bodies; while the outer satellites kept most of their gas and ice, forming larger and less dense bodies. This process mirrored the formation of the major planets.

The majority of information about Jupiter's moons was obtained by the Pioneer 10 and Voyager 1 and 2 missions. Here are the highlights of the data sent back to Earth:

• Io is the most volcanically active body in the Solar System. The gravity of Jupiter pulls on Io, powering the numerous active sulphurous volcanoes. Io's ejected material forms a gaseous sulfur band around Jupiter. There is also a tube of electrical energy connecting Io to Jupiter—more powerful than all the electricity produced by all of Earth's power stations.

• Europa seems to be covered with a 60 mile-thick layer of ice. Gravitational heating may maintain a huge liquid ocean of water under its icy crust.

• Ganymede shows two different types of crust. One appears to be a very old, heavily cratered surface. The other is a grooved surface where the craters have been modified, perhaps by the movement of Ganymede's crust.

• Callisto is the most cratered of the four Galilean satellites. Its icy surface is relatively unchanged since **planetesimals,** leftover chunks of rock from the formation of the planets, bombarded it almost 3.5 billion years ago. (See Chapter 8 for more information on the early Solar System's intense bombardment.)

• The other small moons of Jupiter are irregular bodies of rock that resemble asteroids ranging in size from 70 miles to 4 miles in diameter. Some of these smaller moons were probably captured from the asteroid belt.

The Galileo Mission

The Pioneer and Voyager missions provided astronomers with a brief flyby view of Jupiter and its moons. A much more ambitious mission, called the Galileo, was launched from the Earth on October 18, 1989. Named for the famed discoverer of Jupiter's four major moons, the mission is designed to take detailed, low-altitude pictures of Jupiter and its larger moons.

The Galileo was launched on a very indirect route to Jupiter in order to economize on fuel sending it one pass around Venus and two passes around the Earth to gain momentum. The spacecraft will arrive at Jupiter in December 1995. Much can go wrong during such a long trip, but mission specialists are cautiously optimistic. When Galileo reaches Jupiter, it will circle the planet 11 times in 22 months. The most exciting part of the mission will be the release of a small sensing probe that will descend through the thick Jovian atmosphere. The 6 instruments on the probe will constantly transmit data back to Earth until it is finally crushed by the extreme atmospheric pressures and temperatures of the giant planet.

Saturn

Planetary Observations

To the naked eye, Saturn is a dull yellow color and one of the brightest objects in the sky. At its brightest, it shines at a just less-than-zero magnitude; at its dimmest, it is around 1st magnitude. Because of its greater distance from the Sun, it moves at a much slower rate than Jupiter, completing a revolution around the Sun in just under 30 years.

In 1610, when Galileo first viewed Saturn through a crude telescope, he noticed that the planet appeared to have "ears" or "handles." Later in the century the Italian-French astronomer Jean Dominique Cassini (1625–1712) was able to separate these elongations, describing a system of rings surrounding the planet. Today's observer with a small telescope can see the very same rings and several other features associated with Saturn:

• The planet is oblate.

• Though they are less pronounced than those on Jupiter, Saturn has belts and zones.

• There are light spots in Saturn's clouds, with some of the spots developing into massive storms in the planet's northern hemisphere during its local summer. The last Great White Spot erupted in the fall of 1990 and lasted for more than a month. Similar outbreaks were recorded in 1876, 1903, 1933, and 1960—an almost regular 30-year period.

- As the planet moves in its orbit, bright, visible rings change their angle, as seen from Earth (Figure 5.2). The gap between the two brighter rings, called the Cassini Division, can be seen under favorable conditions with a small telescope.
- The largest satellite, Titan, and the next largest, Iapetus, are easy to observe. Three other satellites (Rhea, Tethys, and Dione) can be seen with a small telescope under excellent seeing conditions.

Rotation and Revolution

Like the other giant planets, Saturn rotates at high speeds, completing one rotation in approximately 10½ hours. (The values of rotation all refer to the outer atmosphere's cloud tops—the only surface visible to observers.) This period of rotation is not the same for the whole planet. The spots at the equator spin around the planet in 10 hours, 14 minutes; while spots at 60°N complete a rotation in 10 hours, 40 minutes. It is unknown if these are the only two basic periods of rotation. The Voyager missions discovered that the equatorial winds of Saturn whip around the planet at 1,000 miles per hour in 10 hours, and 15 minutes. This is faster than the rotation of the planet's interior by 24 minutes, according to theoretical calculations made by considering Saturn's oblateness and internal structure.

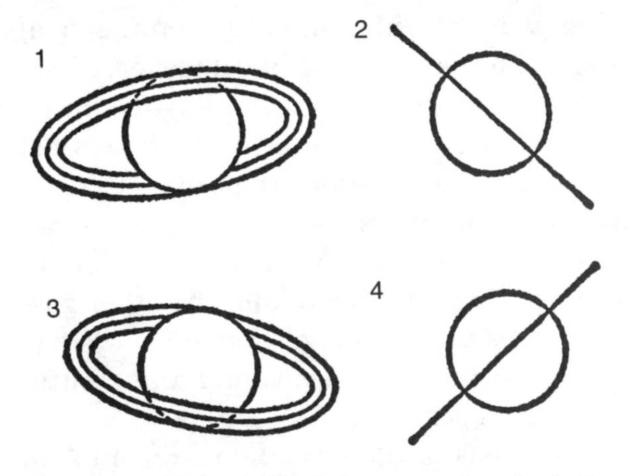

Fig. 5.2. Four positions of Saturn's rings as seen from the Earth.

Saturn orbits at a mean distance of 886,700,000 miles from the Sun and at an average velocity of 6 miles per second. It completes one orbit in just under 29½ years. Saturn's axis is tilted to the plane of the ecliptic at an angle of less than 26½°, or just slightly more than the tilt of the Earth's axis. Its orbit is nearly circular with only a 0.5 percent deviation.

The Structure of Saturn

The structure of Saturn greatly resembles that of Jupiter. The atmosphere is approximately 30,000 miles deep and is made up primarily of hydrogen gas and helium gas with smaller amounts of ammonia and methane. Most likely, the interior of the planet is a rock-ice core 18,000 miles in diameter. It is surrounded by a 6,000-mile-thick layer of liquid, metallic hydrogen with pressures equal to three million Earth atmospheres. The outermost layer is a mixture of liquid hydrogen and helium and is about 16,800 miles thick.

The 1990 Great White Spot and similar storms are probably fresh crystals of ammonia ice. It is thought that the crystals are swept up into the upper atmosphere by the heat from solar radiation during the northern hemisphere's local summer. In addition, Saturn, like Jupiter, has an internal source of heat, which generates the energy that is always driving its weather systems. Planetary scientists were interested in observing the Great White Spot because it helped them track the prevailing winds of Saturn. Normally, it is difficult to do this because of the usual bland appearance of Saturn's markings.

The Rings of Saturn

Saturn is most known for its system of beautiful rings. One generation after Galileo observed bulges on either side of the planet, the Dutch astronomer Christiaan Huygens (1629–1695) conceived the idea that a solid, rigid, and relatively thick ring revolved around Saturn. Huygen's solid, single ring was found to be a system of rings in 1676 when Jean Dominique Cassini (1625–1712) discovered the gap between the two brightest rings that now bears his name.

In the nineteenth century a third translucent inner ring was discovered. The three rings bore the names A, B, and C: Ring A was the outermost, ring B was the middle and brightest, and ring C was the innermost translucent, crepe ring. An innermost D ring and an outermost E ring were also suspected, based on pre-Voyager telescopic observations. (For the classical view of Saturn's rings, see Figure 5.3.)

Until the Voyager missions of 1977, all knowledge of the rings was based on less than ideal, Earth-based observations. As-

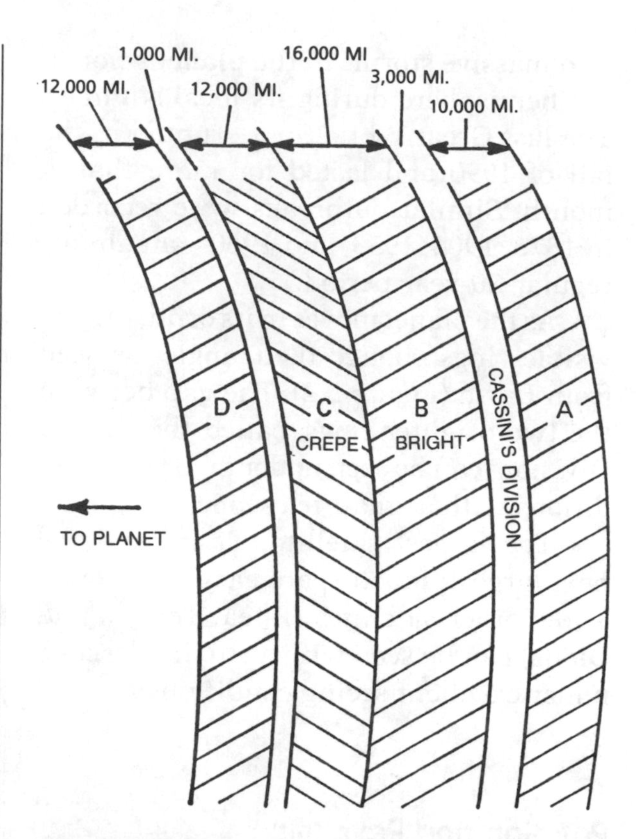

Fig. 5.3. The rings of Saturn. A top view of Saturn shows clearly the four rings as well as the gaps between them. The widths of the rings are 10,000, 16,000, 12,000, and 12,000 miles, respectively. The B ring is the brightest, followed by A, C, and D, in order of decreasing brightness. The 3,000-mile gap between the A ring and the B ring is known as Cassini's division.

tronomers believed that each planetary ring was a wide, smooth, symmetrical sheet that never changed. However, close-up photographs from the Voyager missions shattered this description. Instead of the stable and old-as-the-Solar System rings, astonomers have discovered a dynamic and constantly changing variety of objects ringing Saturn (and the other gas-giants, as well). In fact, they have discovered that the rings are young and continually changing.

The rings of Saturn now run from A to G, with the narrow, braided F ring defining the outermost edge of the system. The combined mass of Saturn's rings is equal to Saturn's small moon Mimas (with a diameter of approximately 200 miles). The rings are not very thick, with estimates of 3 miles for the main rings and less than 100 feet in the other rings. The individual particles in the rings seem to be made mostly of ice and range in size from dust and golf balls to houses.

The rings orbit Saturn almost directly above the planet's equator in a nearly circular path. Current belief is that collisions among the many individual particles quickly flatten and circularize the ring into the flat, thin shape we see today.

But what keeps the rings from spreading and the individual particles from eventually dispersing into space? The Voyager missions were able to detect small moons near and in the rings. These tiny satellites seem to shepherd the particles within the rings—much as a shepherd keeps individual members of a flock together. The gravity of these small ringmoons nudge any straying particle back toward the ring if it attempts to leave the system (Figure 5.4).

By carefully analyzing all the recent data from the Voyager missions, some scientists have concluded that Saturn's rings (most certainly the A ring and the rings around Jupiter, Uranus, and Neptune) are only 10 million to 100 million years old. The following discussion poses some of the possible theories on the formation of Saturn's rings and, in general, other planetary rings.

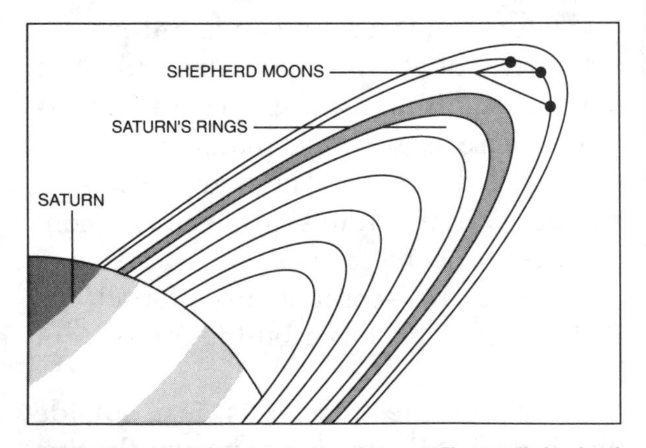

Fig. 5.4. Saturn's rings and shepherd moons. The small shepherd moons are believed to help maintain the shape of Saturn's rings.

The Formation of Ring Systems

Astronomers believe today that there are three basic ways in which rings may have formed:

1. Just as the Sun and its planets formed from a collapsing, cooling solar nebula, each planet had its own condensing nebula that formed a system of moons. Because of the strong tidal pulls of the parent planet, the particles closest could not condense. Only this leftover debris, what we see today as rings, remains from the original creation of the planet and its moons. After the Voyager mission, this theory has become less popular especially since the dynamics of the ring system are so complex.

2. In the nineteenth century, Edouard Roche (1820–1883) proposed another theory. He stated that if a moon were to come too close to its parent planet, the moon passes a limit (the Roche limit) where the tidal forces of the planet tear the moon apart. The resulting debris quickly collapses to form a flat ring. The problem with this theory is that it would take an unusual situation (such as the gravitational pull of a large body) to move a moon into the Roche limit. When only Saturn was thought to have a ring system, this explanation was possible; but with the discovery of other ring systems, it is hard to believe that this happened to each of the ringed planets.

3. A third theory has been recently proposed by Eugene Shoemaker of the United States Geological Survey, based on studies of the cratered surfaces of Saturn's, and the other gas-giants', moons. In particular, Saturn's moon Mimas bears a large impact crater that nearly destroyed the moon. Shoemaker suggests that some smaller moons may have already been destroyed. If a small moon inside the Roche limit were smashed to bits by a large meteoroid, the debris would quickly form a ring. Perhaps the present ringmoons are the larger pieces of this debris left over from the earlier destruction of satellites. And since this has probably been a continuous process for millions of years, the ringmoons may be a source for the younger rings around the gas-giants.

The Satellites of Saturn

Until the Voyager flyby missions, Saturn was believed to have 10 satellites. Today the number has risen to 23—and more are suspected to exist. In general, the moons of Saturn form a regular system around the planet; that is, except for Phoebe, they move counterclockwise, in near circular orbit directly above Saturn's equator. Except for Phoebe and possibly Hyperion, the moons have periods of rotation that match their periods of revolution, so that the same face always points toward Saturn. The moons range in size from that of an average asteroid to larger than the planet Mercury. Only one, Titan, has an atmosphere. This large moon is the densest of all the moons and is half rock and half ice; the other moons seem to be primarily ice, as indicated by their very low densities.

Close-up photographs from the Voyager missions gave astronomers a wealth of new information about many of the moons of Saturn:

- The nearest moon was discovered by Voyager 1 in 1980 and given the designation 1980S28. Its orbit lies just outside the outer edge of the A ring at a distance of 63,000 miles from the center of Saturn.
- The most distant moon is Phoebe, orbiting at a distance of slightly more than 6 million miles from Saturn. This satellite is noteworthy for three reasons. First, it is the only satellite of Saturn to orbit in a **retrograde** motion; that is, it orbits clockwise as seen from above Saturn's north pole. Next, its orbit is tilted toward Saturn's equator at an incredible 150°. Finally, it is most likely the only satellite not to show the same face always to Saturn. Since Phoebe is so unique, scientists believe it may be a captured asteroid or some primordial body. It may have been pulled toward the outer solar system by the gravity of the growing Jupiter and then settled into an orbit around Saturn.
- Titan, the largest moon of Saturn has a diameter of about 3,500 miles and is the only one with an atmosphere. In fact, its atmosphere is so thick that it hides the surface of the moon. There are strong indications that Titan's atmosphere contains **organic** molecules, such as ethane and methane—components found in amino acids, the basic building blocks of life.
- The satellite Enceladus, just outside the very faint E ring, is the most reflective body in the Solar System and shows the

smoothest surface of all of Saturn's satellites. Such a surface may be a result of relatively recent geological activity pushing fresh ice water to its surface.

• Iapetus shows the greatest variation in brightness. This next-to-the-outermost satellite, with a diameter of about 900 miles, is five times brighter when on one side of Saturn than on the other side. Voyager 2 photographs clearly show that Iapetus' trailing hemisphere (the side facing backward with respect to its orbital motion) is bright, with an albedo of 50 percent; while the opposite, or leading hemisphere, is very dark, with an albedo of only 3 to 5 percent. One theory, proposed in 1974, stated that dark matter fell from space (possibly from the surface of Phoebe) onto the surface of Iapetus. However, current belief is that the dark material is part of a type of lava made of ammonia, ice, and perhaps rich, organic material (such as found in carbonaceous meteoroids) that once erupted and spread over the surface of one hemisphere.

• Hyperion is the next moon inward from Iapetus and is one of the largest irregularly-shaped bodies in the Solar System. It is about 240 miles long and 132 miles wide.

• The two tiny moons that lie between the F and G rings are called 1980S1 and 1980S3. They are unique because these two moons are co-orbital; that is, they both occupy the same orbit with slightly differing speeds. As a result, they literally change places every four years.

The Cassini Mission

In the mid-1990s, the remote space probe Cassini will be launched from Earth. Six years later it will arrive at Saturn to begin the first of 36 planned orbits of the ringed planet. Its orbital flight path will allow the spacecraft to study not only Saturn's equatorial regions but higher latitudes as well. There will also be several close encounters with Saturn's moons and rings.

However, the most dramatic highlight of the mission will be the release of a probe into the atmosphere of Titan, Saturn's largest moon. Named Huygens after the seventeenth-century discoverer of Titan, the probe will descend by parachute, to analyze the moon's atmosphere and search for complex organic molecules. If the probe survives the landing, it could give astronomers valuable information about Titan's surface as well.

Uranus

Uranus, the seventh planet, orbiting at a mean distance of 1.782 billion miles from the Sun, greatly resembles the other gas giants. The planet Uranus was accidently discovered by the English astronomer Sir William Herschel (1738–1822) on March 13, 1781, using his handmade 7-inch reflecting telescope. It was the first modern planet ever found. In Herschel's telescope, the planet appeared as a very small disk, little different than the light of a typical star. The difference was enough to lead Herschel to believe that he had found something other than a star. His calculations on the movement of Uranus confirmed that the new find was a planet orbiting the Sun at a distance of 19 AU.

Earth-based observations and the later data from the Voyager missions have provided astronomers with the following picture of the blue-green planet:

- Like the other gas-giant planets, Uranus has an atmosphere made of primarily hydrogen, methane, and traces of ammonia. The planet appears blue because the methane absorbs reddish light.
- The atmosphere is banded like the other gas-giants, but has the least features. There is a thick smog layer in its stratosphere.
- The rotation of the planet, rather than weather, dominates the flow of atmospheric gases and causes the clouds to move.
- It is the only gas-giant that lacks a significant internal heat source. In fact, Uranus is about the same temperature as Neptune (−355° F), despite Neptune's being 50 percent farther from the Sun.
- Like the other gas-giants, the planet has a system of rings and numerous satellites.

Revolution and Rotation

From its distant position in the outer Solar System, Uranus is bathed in an icy twilight from a pinpoint Sun. This is the far region of the Solar System where comets like Halley turn around in their orbit and begin the long trip back toward the inner Solar System. It takes Uranus 84 years to complete one orbit around the Sun at an orbital speed of 4.2 miles per second.

One of the most striking characteristics of Uranus is the tilt of its axis. The axis is inclined an incredible 98° to the plane of its orbit, making Uranus literally on its side. The planet also spins backward, or retrograde, in contrast to all the other planets, which rotate in a counterclockwise direction. Uranus completes one rotation around its axis in 10 hours and 49 minutes. For an illustration of the effects of such an inclination, see Figures 5.5 and 5.6.

One theory states that Uranus' unique tilt was caused by a tremendous impact from some large celestial body early in its life. This collision is also thought to have been responsible for disrupting Uranus' internal heat source.

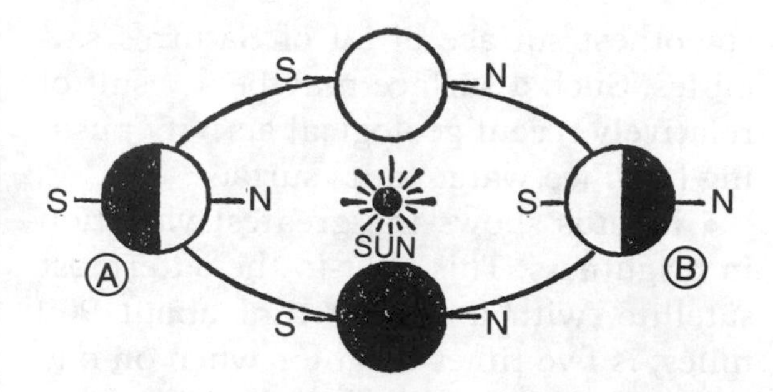

Fig. 5.5. Effect of large (hypothetical) inclination of Earth's axis. If the Earth's axis were inclined by 90° (and not 23½° as in reality) to the axis of its orbit, then at one time during the year the North Pole would be the warmest place on the Earth (position A). Six months later sunlight would shine most directly at the South Pole and that would be the warmest place on Earth (position B).

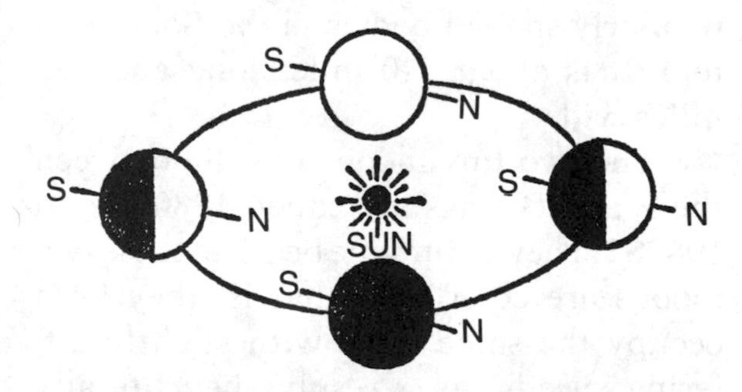

Fig. 5.6. If the Earth's axis were rotated still further from its present inclination of 23½° to, say, 98°, the poles would interchange their identities. The North Pole would now be below the plane of the ecliptic and would therefore be known as the South Pole. The United States would be in the southern hemisphere. An observer looking down at the new North Pole (the former South Pole) would see the Earth rotating clockwise. Clockwise rotation is the normal direction for the Earth's South Pole.

The Structure of Uranus

Most likely, Uranus has a rocky core covered by a mantle of liquid water, methane, and ammonia. This is surrounded by an immense atmosphere of hydrogen and helium gas. Uranus is 67 times larger than the Earth; yet only 14.6 Earths are equal in weight to Uranus. This indicates a low density (0.23 times that of Earth)—typical for all four gas-giants.

Data from the Voyager mission also revealed, for the first time, that Uranus has a strong magnetic field that is greatly offset from its rotational axis. Such an offset was never seen before, and scientists are calling this discovery a totally new class of magnetic field. Voyager also found this type of magnetic field in Neptune (Figure 5.7). Scientists believe that the magnetic fields of Uranus and Neptune are generated in the liquid mantles, which are electrically conductive, rather than in their cores, as is the case with the terrestrial planets.

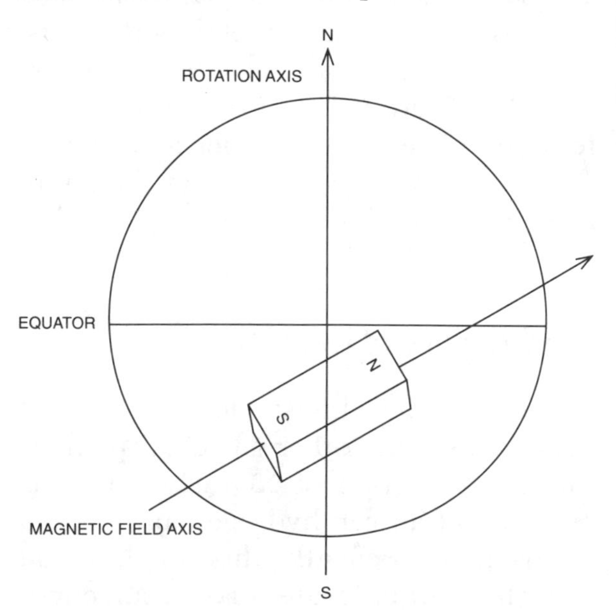

Fig. 5.7. Uranus' magnetic axis. The magnetic axis of Uranus is tilted far from its rotational axis.

The Rings of Uranus

The rings of Uranus were first found in March of 1977 from observations made when the planet passed between the Earth and several stars. To the terrestrial observer, Uranus appeared to cover, or **occult,** these several stars. Observations made from an airborne observatory found that something very close to Uranus also caused the stars to blink out briefly. These findings gave strong evidence for a system of rings around Uranus. Between 1981 and 1985, telescopic observations showed evidence for a system of rings around Neptune. When Voyager arrived at Uranus in 1986, it found a total of 9 narrow rings.

Uranus' rings are not as wide or as spectacular as the rings around Saturn. The Uranian ring system puzzles astronomers because they have not spread into wider patterns. Research into the dynamics of this system continues. In addition, the rings of Uranus are not as bright as Saturn's and are made of an extremely dark material. Combined with the feeble light from the distant Sun, it is impossible to see the rings—even if you were close to the planet! They are visible only on the long exposure photographs taken by Voyager's sensitive camera.

The Moons of Uranus

Uranus has the typical large number of satellites of a gas-giant planet. The five largest, from the closest to the farthest from Uranus, are Miranda, Ariel, Umbriel, Titania, and Oberon—all were known before the Voyager mission. Images sent back from Voyager 2 revealed 10 more small moons and, like the rings, all seem to be made of a very dark material.

Miranda is the most interesting moon of Uranus because of its various types of surfaces. Much of the satellite shows old cratered terrain, but it also has large regions that resemble scoops of marble-fudge ice cream. No one is sure what these features indicate. However, the chevron formations may be the result of relatively recent meltings of the surface caused by massive impacts. Laboratory experiments have demonstrated that if a body made of rock and ice were smashed, then reassembled, similar features would result. Perhaps Miranda has been torn apart and rebuilt several times in its history.

Neptune

Neptune is the eighth planet from the Sun. It is invisible to the naked eye because of its great distance from the Earth. Telescopes reveal a small bluish-green disk that is equal in brightness to an 8th magnitude star like Sirius B—the small companion to Sirius, the brightest star in the northern hemisphere's winter sky.

Interestingly, Neptune was first discovered by mathematical computations, then by actual observation. Astronomers first suspected that there was another planet because Uranus moved strangely in its orbit. By 1845, the difference in the mathematical position and the actual observed position of Uranus strongly suggested that an unknown planet was orbiting beyond Uranus.

In the mid-1840s, John Couch Adams (1819–1892), an undergraduate at Cambridge University in England, and a young French mathematician, Urbain Leverrier (1811–1877), independently computed the orbit of this hypothetical planet—both arrived at the same results. Leverrier immediately sent his calculations to J.G. Galle at the Berlin Observatory. Within an hour after the telescopic search began, the planet was found in the exact position indicated by the mathematical calculations.

Revolution and Rotation

Neptune is far from the Sun at a mean distance of 2.794 billion miles. It takes the planet almost 165 years to complete one orbit around the Sun at an orbital velocity of 3.4 miles per second. The planet's orbit is nearly a perfect circle with only a very small tilt to the plane of the ecliptic.

Studies of the planet's atmosphere reveal that the planet rotates on its axis in 15 hours and 48 minutes, giving Neptune the longest day of all four of the gas-giants. Its axis is tilted at an angle of a little less than 29°. There is very little seasonal difference between its summer and winter, with the mean temperature, at the top of its clouds, a chilly −350° F.

The Structure of Neptune

Neptune is the almost identical twin of Uranus. Like the other gas-giant planets, the major portion of Neptune's mass is made up of a deep hydrogen and helium atmosphere. Beneath this thick cloud cover lies a liquid water mantle surrounding a hot rocky core. Neptune is the densest of all the gas-giants with a mean density of .33 times that of the Earth. It would take 57 Earths to fill the space occupied by Neptune and 17.2 Earths to equal its mass.

Unlike Uranus, Neptune has many visible cloud and storm systems in its atmosphere. The following features were discovered and photographed by the Voyager mission in 1989:

- *The Great Dark Spot* This oval feature is very similar to Jupiter's Great Red Spot. It is a huge storm system in the southern hemisphere that seems to stretch and contract in an eight-day cycle.
- *The Scooter* Not quite a true storm, this white patch seems to be a peculiar cloud feature that moves quickly relative to the larger features around it. Its shape changes from round to square to triangular.
- *The Bright Companion* This white feature is associated with the Dark Spot and resembles The Scooter. It is one of the few cloud features on Neptune visible from Earth.
- *D2* Another Neptunian storm, this feature developed a bright core lasting several days. The bright material rose high in the atmosphere and, in cross-section, probably resembles a terrestrial thunderstorm. Astronomers believe that this type of feature is a way of transporting certain gases from the lower to the upper atmosphere.

The Rings of Neptune

Between 1981 and 1985, astronomers were faced with a puzzle. They had detected a ring system around Neptune during a star occultation; but it seems that the star blinked out only on one side of Neptune. This meant that the rings were partial, or ring arcs, something completely new in a ring system. The scientists waited for details from the Voyager mission. But with all the Voyager data, there is still controversy: Some astronomers see very faint but complete, rings; while others see the ring arcs.

The Moons of Neptune

Before the Voyager encounter, Neptune was known to have two satellites, tiny Nereid and huge Triton. Nereid is interesting because its orbit is highly eccentric; that is, it is seven times farther from Neptune at aphelion than at perihelion. The tiny moon takes 360 days to complete one orbit around Neptune.

Triton is the only major satellite in the Solar System that orbits opposite to the spin of its parent planet. Scientists believe that Triton was once a free object in the Solar System and was captured by Neptune. Voyager's cameras found a great variety of terrain on Triton. There are regions of large, flooded impact craters, rolling terrain with light cratering, and mysterious markings that resemble cantaloupe skin in other areas.

Triton also has a thin atmosphere, less than 2 miles thick. It is primarily nitrogen driven off from its sunlit polar cap and ejected from active geysers in the form of nitrogen snow. Photographs of these nitrogen plumes and dark streaks have given astronomers a picture of the dynamics of Triton's primitive weather system. Scientists speculate that dark organic material may be buried under a crust of transparent ice and that sunlight may be vaporizing nitrogen ice in the buried dark material. The trapped nitrogen vapor would eventually explode through vents in the upper ice, ejecting the dark matter. The surface temperature of Triton is −391° F, making the moon the coldest place in the Solar System. Triton is thought to resemble the planet Pluto. Scientists are now poring over the Voyager data on Triton to try to gain a better understanding of Pluto—the only planet not yet visited by one of our spacecrafts.

Pluto

The ninth and most distant planet known is Pluto, named for the mythological god of the underworld. It was discovered in 1930 and remains somewhat of a mystery to this day. Imagine a world where the Sun is 900 times fainter than it is on Earth and sunlight reflects off plains of frozen methane. In Pluto's sky is a moon named Charon that looks 60 times larger than our Moon—but it never rises or sets. Instead, Charon is locked forever into the same position above the horizon.

Pluto is mysterious for several reasons:

- It is extremely far away, making visual observations very difficult.
- Although certain astronomers predicted a planet to exist beyond Neptune, Pluto's size and composition turned out to be very different from the predictions.
- Its relationship to its moon is unique in the Solar System.
- Its orbital motions are also unique in the Solar System and has led some astronomers to question whether Pluto is a true planet, or perhaps an asteroid.

Pluto's Orbit

On the average, Pluto is about 40 AU from the Sun, or 3.6 billion miles away. Its orbit brings the planet as close as 2.75 billion miles at perihelion, and as far as 4.6 billion miles at aphelion. Because of its extreme distance Pluto takes 248 years to make one revolution around the Sun. Normally, Pluto lies much further away from the Sun than Neptune. However, the large eccentricity of its orbit brings the planet closer to the Sun than Neptune near perihelion. In 1979, Pluto slipped inside the orbit of Neptune and will remain there until 1999. Pluto's orbit has the greatest inclination to the ecliptic of all the planets (17°). This means the two planets will never collide since Pluto is always at least 250 million miles above Neptune's orbit when this crossing takes place (Figure 5.8). This close approach to Neptune and Pluto's size and period of rotation (6.39 days) have led some astronomers to believe that Pluto may have once been a moon of Neptune.

The Composition of Pluto

No one really knows the true composition of Pluto. Planetary scientists usually can determine the composition of a planet by measuring the relationship between the planet's diameter, mass, density, apparent brightness, and albedo. But in the case of Pluto, astronomers have accurate values for only it's mass (found to be 0.0021 Earth masses) and apparent brightness (13.7 visual magnitude). If they could learn the value of just one other characteristic, then all the other values would fall into place.

The best estimates give Pluto a diameter of between 1,548 and 2,118 miles. There-

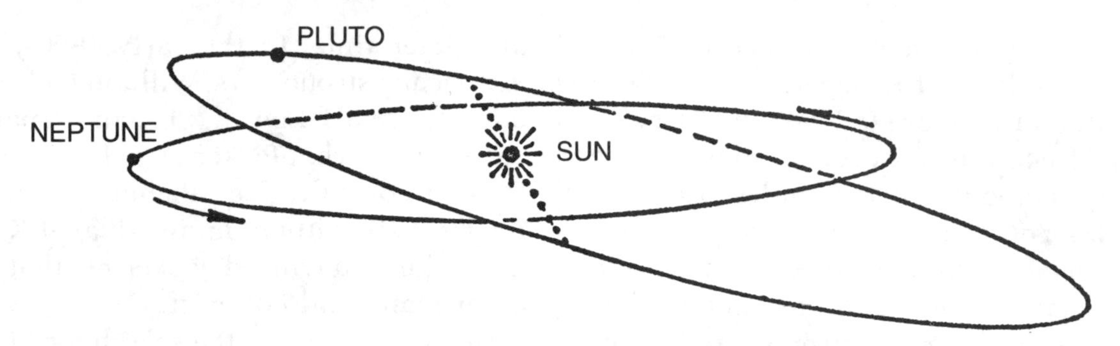

Fig. 5.8. The orbits of Neptune and Pluto. Neptune and all the other planets except Pluto move in orbits that are only slightly inclined to the ecliptic. The large inclination of Pluto's orbit as compared with Neptune's is shown here.

fore, the density of Pluto lies between 10 and 20 percent of the Earth's. This gives significant clues to scientists about Pluto's composition. Pluto most likely has a small rocky core surrounded by a thick mantle of water ice. Covering this mantle is a thin envelope of frozen methane ice (Figure 5.9), probably permanent as the surface temperature ranges between −390° and −350° F. There are possible outcroppings of rock (small mountains could exist as well), and there should be an abundance of craters. Pluto's small size and rocky core qualify it to be called a terrestrial planet. Yet, most current thought states that Pluto is a planetesimal.

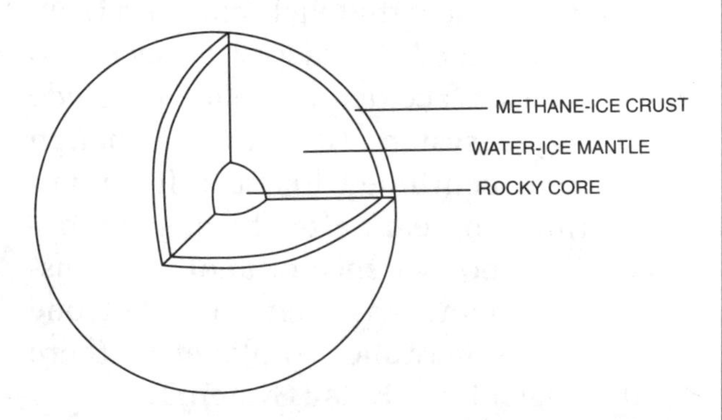

Fig. 5.9. Pluto's interior. A cutaway view of Pluto's possible internal structure.

Pluto's Atmosphere

In 1976, spectroscopic study showed methane ice on the surface of Pluto. Some astronomers believe that this ice may form a thin methane atmosphere when Pluto's orbit nears the Sun. Unfortunately, it is very difficult to tell the difference between methane ice and methane gas from the Earth. Other scientists have suggested that there must be another heavier gas—perhaps argon or nitrogen—or else Pluto's weak gravity, about $\frac{1}{12}$ of Earth's gravity, could not have held on to the atmosphere. So far, there has been no confirmation of anything besides methane.

Pluto's Moon

In 1978, astronomer James Christy discovered Pluto's moon, Charon, when he noticed elongations of Pluto's disk on several photographic plates. Its existence was not officially recognized until 1985, when Pluto and Charon begin a series of mutual eclipses. These motions of the two bodies convinced even the most skeptical observers that it really was a moon and not some high mountain on the surface of Pluto. Charon is difficult to separate from Pluto because it is so close to its parent planet—only 11,640 miles away. (In contrast, the Moon is about 250,000 miles from the Earth.) Yet, Charon takes a leisurely six days to circle Pluto. (A moon that close to the Earth would speed around it once every seven hours!) Astronomers believe that the true mass of Pluto must be much less than originally predicted because Pluto's gravitational pull on Charon is very weak.

Since Charon revolves around Pluto in the same period that Pluto rotates about its own axis, Charon seems to hover over the same region of Pluto. Charon also keeps the same face toward Pluto and vice versa. When the Pluto-Charon system passed in front of a star in 1980, Charon eclipsed this star for 50 seconds. This figure indicated that Charon has a diameter of at least 720 miles, ranking it as the most massive satellite in the Solar System relative to its parent planet. Little else is known about this satellite except that it must resemble Pluto—containing methane, a low density, and extremely cold temperatures.

The Discovery of Pluto

The discovery of Pluto is a testament to patience and a story of misleading predictions. For a long time, it was known that something was pulling on the giant planets Uranus and Neptune and causing them to move slightly in their orbits. In fact, Neptune was discovered by purely mathematical methods and then visually sighted at a later time. In the early 1900s, two American astronomers, William Pickering and Percival Lowell, began separate searches for what Pickering called "Planet O" and Lowell called "Planet X." Lowell died in 1916 without finding Planet X and his failure convinced Pickering that another planet did not exist.

However, in 1929, the search for Planet X resumed at Lowell Observatory near Flagstaff, Arizona, under the direction of twenty-two-year-old Clyde Tombaugh. Although he had no formal training as an astronomer, he had youthful energy, determination, and a photographic memory. He spent two years comparing photographic plates of the sky in order to find any movement that would indicate a planet. On February 18, 1930, he found what he was looking for. Late on March 12, 1930, a formal announcement was made of the discovery of Pluto. In honor of Percival Lowell, the letters "P" and "L" combined were designated as the symbol for Pluto.

But debate has continued about Pluto being the true ninth planet—even decades after it was found. Simply stated, Pluto is really too small to account for the pulling of Uranus and Neptune. In a way, Pluto was not the object that Pickering and Lowell had predicted. Its discovery seems to have been an accidental result of Clyde Tombaugh's systematic search. Although Tombaugh continued his search for another thirteen years after having discovered Pluto, no evidence of another trans-Neptunian planet has been found. Yet, one wonders—is there another planet out there at the edge of our Solar System?

CHAPTER SIX

Minor Bodies of the Solar System

KEY TERMS FOR THIS CHAPTER

alloy
gravitational resonance
meteor
meteor shower
minor planet
radiant

Asteroids

One planet is missing. Theoretically, there should be a planet revolving in orbit between Mars and Jupiter, but no planet has ever been found there. Instead, there are a large number of small bodies, known either as asteroids, **minor planets,** or planetoids. Some have diameters as large as 500 miles; others are smaller than a mile. The first asteroid discovered was Ceres in 1801. The next three, Pallas, Juno, and Vesta, were discovered in 1802, 1804, and 1807, respectively. The number of known asteroids is in the thousands. Their often irregular shapes has led astronomers to believe that they are the fragments of a planet that never formed because of the gravitational pull of neighboring Jupiter.

Theoretical Discovery

As so often happens in astronomy, the asteroids were first discovered in theory, then later actually observed. The theory was based on the Titius-Bode law. In 1766, the German mathematician Johann Titius presented the law to a skeptical astronomical community (though it had been suggested as long ago as 1596 by Johann Kepler). In 1772, German astronomer Johann

Elert Bode (1747–1826) revived and published the idea.

The Titius-Bode law states that the distances to the planets can be mathematically calculated. Try the following:

1. List the planets in order from the Sun.
2. Write the number 4 under each planet.
3. Write products of 0 x 3, 1 x 3, 2 x 3, 4 x 3, 8 x 3, and so on, under each planet in proper order.
4. Add the vertical columns and divide by 10.

Following these instructions, one would arrive at the following:

Merc	Venus	Earth	Mars	?	Jup	Sat
4	4	4	4	4	4	4
0	3	6	12	24	48	96
.4	.7	1.0	1.6	2.8	5.2	10.0

The bottom row of numbers closely corresponds to the true distances of the planets from the Sun when expressed in astronomical units (AU). The true distances are

.39 .72 1.0 1.52 2.8 5.2 9.54

According to this law, there should be a planet at a distance of 2.8 astronomical units from the Sun. A systematic search for the missing planet in the area of 2.8 AU led to the discovery of a vast number of minor planets. The first, Ceres, named for the guardian god of Sicily, was discovered on January 1, 1801, by the Italian astronomer Giuseppe Piazzi (1746–1826). Its distance from the Sun matched closely with the distance obtained from the Titius-Bode law.

Not all the planets fit into the Titius-Bode law. Uranus, Neptune, and Pluto had not been discovered when Bode's rule was published (1772). Uranus fits the rule nicely and was discovered shortly after the rule was published. However, Neptune and Pluto do not conform to this rule.

An explanation of Bode's rule is given in a theory known as "dynamical relaxation." According to this theory, the planets, when first formed, followed an entirely different set of orbits. Reacting to gravitational forces from its neighbors, each planet changed its orbit until the disturbing forces were minimal. The final arrangement that we see today is one that obeys the mathematical expressions similar to the Titius-Bode law.

The Naming of Asteroids

As soon as a minor planet's orbit is established, it is designated by a number—in order of its discovery—followed by a name (for examples, 1 Ceres, 2 Pallas, 3 Juno, and so on). The discoverer of the asteroid usually selects the name. In the beginning, feminine names were taken from mythology. Later, names were taken from Shakespeare's plays and Wagner's operas. Many asteroids bear the names of wives, friends, institutions, famous people (the Beatles!), and even pets.

The Composition of Asteroids

Scientists are especially interested in the composition of asteroids because these minor bodies may hold a key to understanding the early history of the formation of the Solar System. The primordial material that formed much of the Solar System can be found in the asteroids in an almost unchanged form. For example, Ceres is composed mostly of carbonaceous material; while Vesta is almost pure basaltic in its composition. These materials are thought to have played a major role in the origin of the Solar System. In addition, the asteroids may someday be a source of material for mining interests.

Orbits of the Asteroids

The vast majority of the asteroids are found between the orbits of Mars and Jupiter. The perihelia of some asteroids, for example, 433 Eros, lie within the orbit of Mars. Others, known as the Apollo asteroids, have perihelia inside the Earth's orbit. There are a number of asteroids with aphelia outside the orbit of Jupiter. For example, 944 Hidalgo has an aphelion distance of 9.64 AU.

The periods of revolution of the minor planets around the Sun vary greatly between a lower limit of 2 years and an upper limit of 12 years. There are, however, so-called forbidden periods—no asteroid has a period one half, one third, or one quarter the period of Jupiter (11.86 years). This is a direct result of a physical phenomenon known as **gravitational resonance,** where the gravitational attraction of the planet Jupiter constantly adjusts the orbits of the asteroids into certain configurations.

All asteroids move around the Sun in a counterclockwise direction, like the nine major planets. But the inclination of the orbits of the asteroids to the ecliptic varies within wide limits. Many of them orbit very closely to the plane of the ecliptic; others move far below or above the plane of the ecliptic. The orbit of Icarus is shown in Figure 6.1. Its orbit is tilted 21° to that of the Earth's orbit around the Sun.

The Trojan and Near-Earth Asteroids

Trojans, or the Jupiter asteroids, is the name given to two groups of asteroids that follow the same orbit as Jupiter: One group travels 60° ahead of Jupiter, the other 60° behind. The location of these asteroids in relation to the planet agrees with a theoretical solution of the three-body problem formulated by the French mathematician Joseph Louis Lagrange (1736–1813). A detailed explanation of this problem is beyond the present scope of this book.

There are asteroids whose orbits take them close to the Earth's orbit. Several have been recently discovered that pass within less than 10 million miles from the Earth's orbit. Astronomers keep a close eye on these near-Earth asteroids. On October 10, 1937, the small asteroid Hermes came within half a million miles of the Earth. In June 1968, Icarus passed within 3 million miles of the Earth. More recently, in January of 1991, the newly discovered asteroid 1991BA passed within 102,000 miles

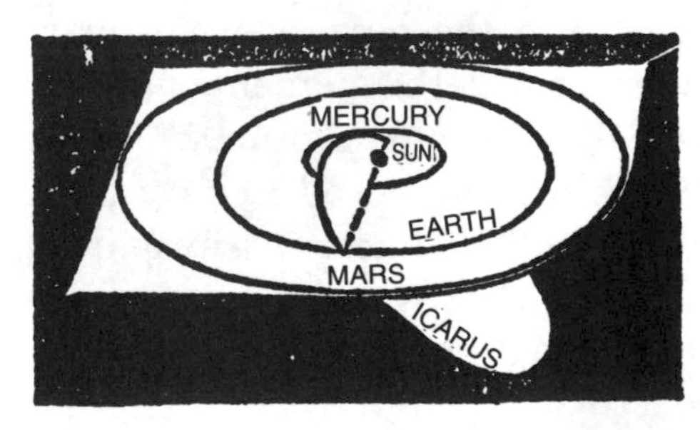

Fig. 6.1. The orbit of Icarus is inclined 21° to the orbit of the Earth around the Sun.

of the Earth. Based on its brightness, 1991BA could not have been more than about 40 feet across, thus making it the smallest asteroid ever observed. Yet, had this object actually struck the Earth, its impact energy could have destroyed a small city.

Comets

The name "comet" is derived from *stellae cometae,* Latin for "long-haired stars." They are perhaps the most remarkable objects in the Solar System primarily because their appearance differs so greatly from what they really are. A bright comet can appear like a large illuminated disk, sometimes the size of the Moon. At times it is visible in the daytime sky. It often sports a wispy tail full of knots and filaments that can stretch for millions of miles. This celestial wanderer seems to move around our Sun like some planet traveling in a very elongated ellipse.

The vast dimensions of some comets, together with their extraordinary brilliance, are no doubt responsible for the many superstitions associated with their appearances. The arrival of a bright comet was believed to be "ominous of the wrath of Heaven, and the harbingers of war and famine, of the dethronement of monarchs, and the dissolution of empires." The comet of 43 B.C. was supposed to have been the soul of Julius Caesar transported to heaven. A comet was associated with the coming of William the Conqueror when it appeared in April of 1066. The comet of 1811–1812, of course, was called "Napoleon's Comet."

The Origin of Comets

Billions of years ago, many planetesimals—small pieces of rock and ice that eventually formed today's planets and moons—may have first been created and then flung far from the Sun to form clouds of comets when the Sun first ignited. In 1950, Dutch-American astronomer Jan Oort proposed that those icy planetesimals were then tossed by the gravitational fields of Uranus and Neptune into orbits that carried them thousands of AUs away, but they remained gravitationally bound to the Sun. Oort calculated that these planetesimals still exist in a spherical shell that surrounds the Solar System. In fact, Oort based his calculations on the orbits of long-period comets, those with periods longer than 200 years.

Oort determined that the comets attracted into highly elliptical orbits by Uranus and Neptune would be pushed into more nearly circular orbits by encounters with passing stars. These encounters would also have the effect of scattering comets above and below the ecliptic plane, creating a spherical cloud of comets rather than a flattened disk. Today, astronomers believe that the Oort Cloud extends from about 20,000 to 100,000 AU (almost 2 light years) from the Sun and contains as many as 2 trillion comets with a mass several times that of the Earth's (Figure 6.2).

For many years, astronomers assumed that all comets observed in the Inner Solar System (that is, crossing the orbits of the planets) came from the distant Oort Cloud. During the 1980's, some comet specialists speculated that the Outer Oort Cloud comets may be outnumbered by another inner cloud that begins about 3,000 AU from the

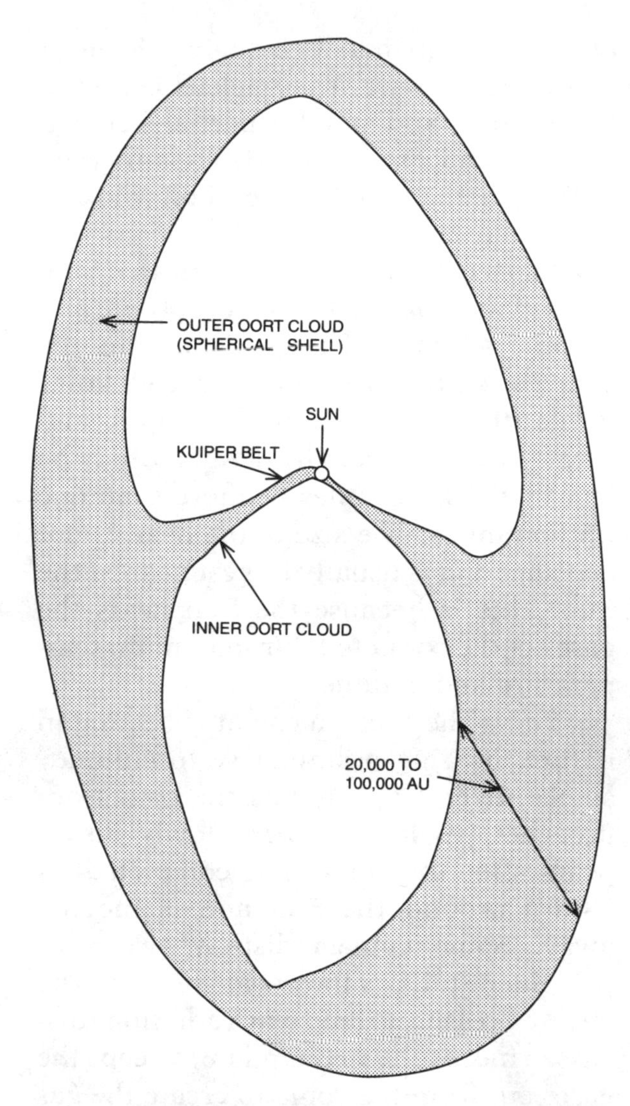

Fig. 6.2. Cross-section of the Oort Clouds. Astronomers believe that as many as 2 trillion comets surround the Sun in a vast spherical shell known as the Oort Clouds.

Sun and continues to the edge of the Outer Oort Cloud (but does not include it). The innermost portion of this Inner Oort Cloud is relatively flattened, with comets extending a few degrees above and below the ecliptic plane. But the Inner Oort Cloud rapidly expands, forming a complete sphere by the time it reaches several thousand AU from the Sun. Unlike the more distant comets of the Outer Oort Cloud, the comets of the Inner Oort Cloud are tightly bound by the gravity of our Sun, so passing stars and other gravitational disturbances outside our Solar System do not upset their orbits enough to send them out of the System.

In addition, most recent studies of short-period comets, such as Halley's Comet, has led astronomers to conclude that there is yet a third cloud of comets just beyond the orbit of Pluto. This belt begins at about 35 or 40 AU from the Sun and may extend as far as several hundred AU. It has been called the "Kuiper Belt" after Gerard Kuiper, who first proposed the possibility of its existence in 1951. The gradual pull of the giant planets over time attracts an occasional comet from the Kuiper Belt toward the Sun (Figure 6.3).

The Structure of a Comet

Comets differ greatly from all other objects in the Solar System. They are quite unique in size, mass, density, and behavior. The tails of some comets extend 50 million miles; some reach a length of 100 million miles. The width and thickness of an entire comet are also of colossal proportions,

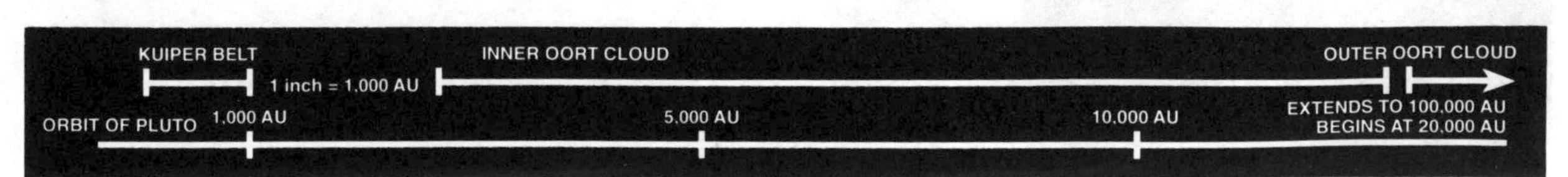

Fig. 6.3. The scale of the distances of the comets surrounding the Sun.

with 50,000 miles a typical figure for either of these dimensions. The tail of the great comet of 1843 was estimated to have been more than 500 million miles long.

Although characteristically very large in volume, the mass of a comet is too little to gravitationally disturb even the smallest satellite on close encounter. The mass of a large comet has been estimated to be one millionth that of the Earth, especially since a comet is largely icy dust and smoke. The tail is virtually transparent—enough to see stars through it.

A comet usually consists of a nucleus, quite small and bright; the coma, a nebulous mass that surrounds the nucleus; and a tail that seems to stream from the coma. The tail is always much dimmer than the head, although there is no sharply defined boundary separating them (Figure 6.4).

• *Nucleus* The nucleus is the central bright spot, with a diameter rarely exceeding a few miles. It differs from the rest of the comet primarily by being the most concentrated part. The nucleus is a mixture of rocky and metallic particles coated with frozen ices of water (H_2O), ammonia (NH_3), methane (CH_4), carbon dioxide (CO_2), and dust.

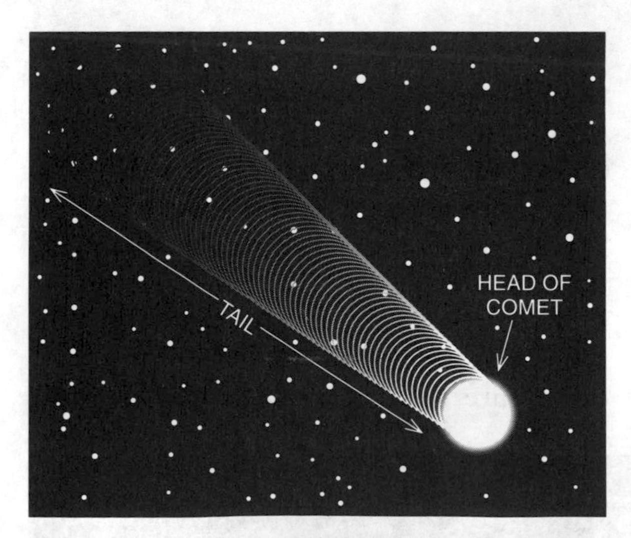

Fig. 6.4. The structure of a comet. The head, or coma, forms the front of the comet with the nucleus the central part of it. The main bulk of the comet is taken up by the tail.

• *Coma* As a nucleus nears the Sun, some of the frozen particles evaporate and become gaseous. These gases form the comet's coma. Most comets have globular heads (the nucleus plus the coma), and greatly vary in size. The diameter may be less than 10,000 miles or more than one million miles. The size of the head is not constant and is usually largest nearest the Sun. This is because the Sun heats the gases of the coma to their maximum temperature and volume.

• *Tail* The tail of a comet is fleeting in nature, bearing a similarity to chimney smoke. It first appears as the comet approaches perihelion, then slowly disappears after the comet has completed its U-turn around the Sun and is moving away. Some comets display two very prominent tails as they near the Sun. The steady pressure from solar radiation (discussed more fully in Chapter 8) sweeps the gases out from the coma to create the gas tail and also pushes out the very fine dust particles to form the dust tail. The solar radiation pushes on the gas and dust in an outward direction, causing the tails to always point away from the Sun. As the comet approaches the Sun, the tails follow the coma; as the coma moves away from the Sun, the tails precede the coma (Figure 6.5). Some comets have even surprised observers by sporting a short antitail, which projects 180° from the direction of the main tails. The origin of this antitail is not yet clearly understood.

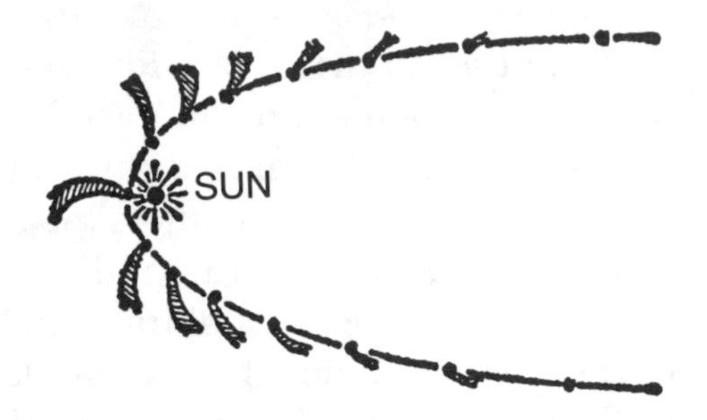

Fig. 6.5. The tail makes its first appearance when the comet is within several astronomical units from the Sun. It is longest at perihelion and disappears when the comet is again further from the Sun. Because of pressure of sunlight, the tail at all times points away from the Sun.

The Discovery of Comets

Very few bright comets have been recorded in all of history. One or two in a lifetime is the average. The last great comet appeared in 1882 and was highly visible for several weeks, although it was observed for months through telescopes after it had faded from naked-eye visibility. No truly spectacular comet has appeared in this century. Two that appeared in 1910 (one, a return of Halley's Comet; the other, the 1910 I comet) were only fairly great comets. Comet West provided a respectable display in 1976 for mid-northern latitude observers; while Comet Kohoutek in 1974 fell far below expectations of brightness and size. The most recent return of Comet Halley in 1986 was also less than spectacular, as was the appearance of Comet Austin in 1990. In the summer of 1990, Comet Levy proved to be quite prominent in moderate telescopes for several months.

Between five and ten new comets are discovered every year, most of them too faint to be seen with the naked eye. Since the middle of the seventeenth century, when the first successful calculations were made concerning the true nature of cometary motions, the orbits of more than 700 comets have been reliably determined. Some of these comets, such as Encke's Comet, have very short periods and return to round the Sun every few years. Others, such as Halley's Comet, take closer to one hundred years to make their round trip; while some may not return for thousands of years. Some never return at all. They make one loop around the Sun and then return to interstellar space (Figure 6.6).

Encounters with one or more of the major planets (especially Jupiter) will often

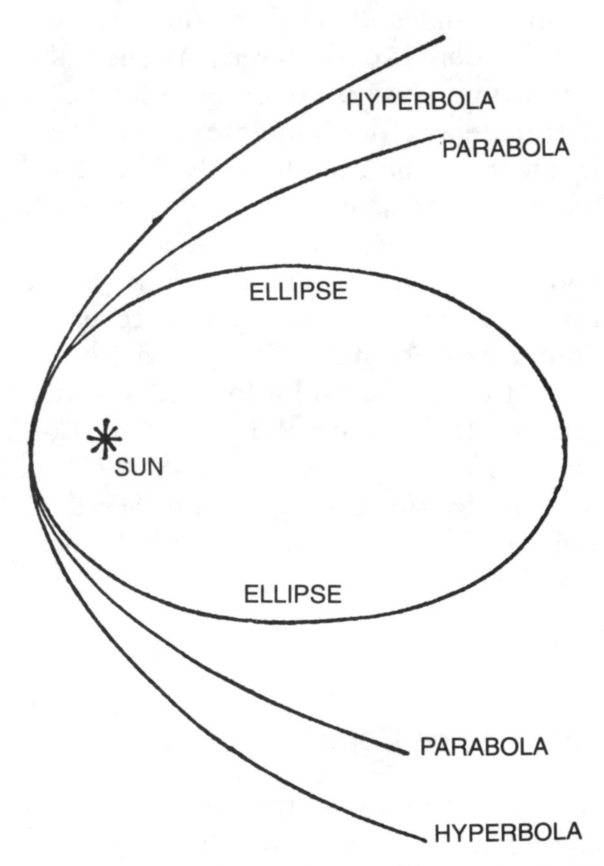

Fig. 6.6. Possible orbits of a comet. An ellipse is a closed orbit; comets moving on it return repeatedly to perihelion and can be seen periodically. Both the parabola and hyperbola are open orbits; the two ends of each never meet. Any object moving in such an orbit makes only one appearance at perihelion and then disappears into deep space.

alter the orbit of a comet. The comet may be kicked further out from the Sun, or it may be nudged closer toward the Sun. Consequently, astronomers have lost comets, only to rediscover some of them at a later date with completely new orbits!

More and more frequently, amateur astronomers have been taking the lead in the discovery of new comets. These amateurs have the freedom unavailable to professionals to scan the heavens in wide swaths. Most professional astronomers spend their precious time and funding on very specific and narrow research projects. They cannot roam the skies at will like the amateur. This freedom to cover large portions of the sky, with a good measure of luck, allows the amateur comet-hunter to discover new members of the comet family. Therefore, many of the discovered comets bear such amateurs' names as West, Austin, Levy, and Ikeya-Seki (hyphenated names mean that there was more than one discoverer). Comets are formally designated by the year of discovery and a lower-case letter to indicate the order of discovery for that given year. For example, Ikeya-Seki 1967n was the fourteenth comet discovered in 1967; while Comet Austin bears the designation 1989c.

Research on Comets

Ground-based research on comets has become a much more active field since the reappearance of Comet Halley in 1986. As a result, there is a more precise understanding of the mechanisms that form and place comets in the outer reaches of the Solar System. Astronomers are interested in comets for two main reasons. First, comets may be made of the same material that originally formed our Sun and the rest of the Solar System. Second, comets may be made of the same material that still exists in the star-forming nebulae of interstellar space. Interstellar molecules can stick to the surface of a comet, coating it with a thin layer of interstellar material. Frozen in place by the deep cold (−459° F) of the Oort Clouds, this material may reveal to astronomers some of the earliest processes that eventually led to the formation of our Solar System.

Numerous spacecraft also have been, and will be, used to study comets. In 1986, the European Space Agency's Giotto probe passed within 360 miles of Comet Halley's nucleus—one of three craft sent to the comet from various countries. In particular, the Giotto made a series of photographs revealing the presence of several types of dust. One type of dust had a mixture of elements in the same proportions as that found in the stars of our galaxy. (The European Space Agency planned to reactivate Giotto for a flyby of Comet Grigg-Skjellerup in 1992.)

Other teams of astronomers want to use the Earth-orbiting Hubble Space Telescope to make high-resolution studies of comets as they pass through the inner Solar System. Researchers are also hoping to one day launch a craft that will fly in formation with a comet, photographing the nucleus, recording temperatures, and analyzing the comet's surface. Scientists hope to fire a probe into the nucleus of a comet resulting in the first direct analysis of material from a comet. The most ambitious plan of all calls for an attempt to scoop some material from the surface of a comet and return it to the Earth for study in a ground-based laboratory.

The Fate of Comets

Each time a comet passes close to the Sun, some of its mass warms enough to vaporize and form a tail. After repeated perihelion passages, the comet uses up all of its volatile elements, no longer glows, and becomes a swarm of dark debris.

In addition to running out of the material that forms the head and tail, a comet may also split into two or more pieces during a perihelion passage. When passing close to one of the major planets, the planet's gravity creates heavy tides within the comet's nucleus. These tides are mainly responsible for the splitting. Roche's limit (see Chapter 5, The Formation of Ring Systems) basically indicates that a comet passing within 90 million miles of the Sun, or 9 million miles from Jupiter, or 2 million miles from Earth, should disintegrate as a result of the tides produced within it. Observations verify this prediction: The 1947 XIV comet passed within 10 million miles of the Sun and was broken in half.

As was noted before, comets may also be only one-time visitors whose open-ended orbits take them back into interstellar space, never to return. Or they may be victims of planetary cue-ball and get knocked completely out of the Solar System or into the Sun. New comets, however, are traveling continuously toward the Sun from the Oort Clouds and the Kuiper Belt. One theory states that our Sun has a small, dim companion star at a distance of one or two light-years. This star, dubbed "Nemesis," may be responsible for gravitationally pushing many of the Oort Cloud comets toward the Sun. No clear visual evidence has shown that such a star actually exists. Perhaps more distant events—such as the explosion of massive stars or the ripples of gravity from giant clouds of molecular hydrogen occasionally sweeping through our galaxy—may also give the necessary push to bring comets in from the outer fringes of the Solar System and provide us with one of nature's more brilliant spectacles.

Meteoroids

Meteoroids are tiny solid objects, mostly the size of sand particles, traveling through space often along the same orbits as comets. A study of their locations and motions indicates that meteoroids are the remains of comets that have lost a great deal of their mass on repeated passages near the Sun. Soon after the death of a comet, with the gravitational attraction of the remaining mass too weak to keep the particles together, the particles form a closely packed group that has been aptly called a "flying gravel pile." Such a group is known as a swarm. With time, there is a great deal of scattering of the particles both along the elliptical orbit and sideways. An elongated pile of such particles, which may extend all around the orbit, is known as a *stream*. Another theory notes that some meteorites follow counterclockwise orbits of low inclinations, suggesting that some meteoroids were once members of the planetoid population in the asteroid belt and not remnants of comets. As the asteroids collide, small fragments break away from the belt and can, over thousands to millions of years, eventually reach the Earth.

Meteor Showers and Shooting Stars

Compact swarms or streams of meteoroids produce meteor showers (see in this chapter, The Frequency of Observable Meteors) that can be observed on certain nights of the year, while scattered streams are responsible for the random meteors that can be seen on any dark, clear night (Figures 6.7 and 6.8). The Earth, moving along its orbit, is continuously colliding with many of these scattered, solid particles, the vast majority of which do not survive this encounter. Upon entering the Earth's atmosphere at a fairly great speed, typically, 20 miles per second, these meteoroids are completely burned up by the heat produced by the compression of the air in front of them and by the friction between the air and the meteoroid surfaces. Meteoroids are first visible from the Earth at heights of 60 to 90 miles; most vanish at heights of 30 to 50 miles.

The creation of the light that results

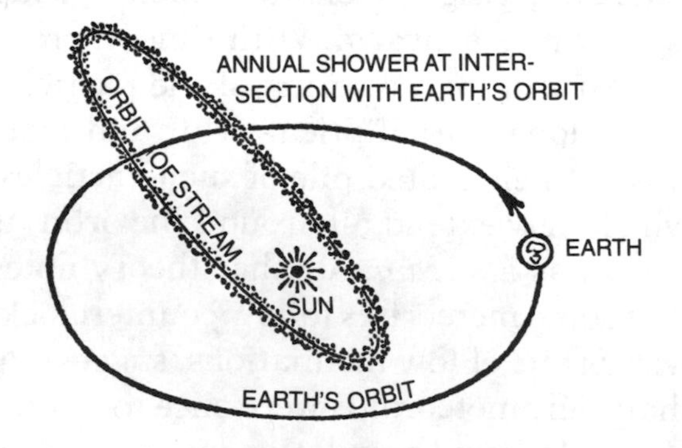

Fig. 6.7. Meteor showers due to a stream. These showers are annual events. They can be seen each time the Earth is close to a stream of meteoroids.

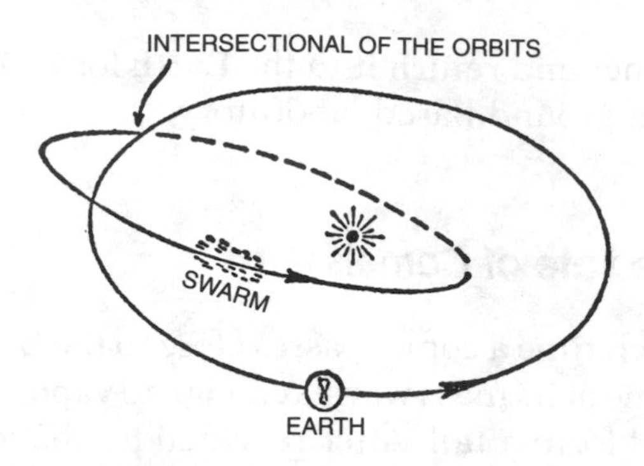

Fig. 6.8. Meteor showers due to a swarm. These showers are much more spectacular. They occur every time the Earth and the swarm are simultaneously at the point of intersection of their orbits. For some swarms this may be every few hundred years.

from a meteoroid's entry into the Earth's atmosphere is called a **meteor,** or shooting star. The light seen by the observer is caused by the collision between atoms knocked off the meteoroids and atoms of hot air. The volume occupied by these colliding atoms is in the form of a tube, where the length is the length of the streak and the cross-section is a circle having a diameter 10 feet or more. The majority of shooting stars are nearly invisible grains of rock and iron that plunge through the Earth's atmosphere. However, if the meteoroid is the size of a basketball or larger, its entry into the atmosphere creates a spectacular sight called a "fireball."

Shooting stars are quite common. On a clear night, five or more per hour can be seen from any single point on Earth. The number all over the Earth's surface on a given night is estimated to be more than 20 million. The number of fainter meteors that can be seen only with the aid of a telescope is thought to be between 5 and 10 thousand million per night. Every day, the dust resulting from these fiery collisions adds over 50 tons of mass to the Earth.

Meteorites

Occasionally, a large meteoroid collides with the Earth's atmosphere, survives, in part, and lands on the surface. Such a survivor is called a "meteorite." Meteorites can be seen on exhibit in various museums, with many of the fragments several feet in each dimension. The Earth has been hit by truly gigantic meteorites. One meteorite, weighing about 300,000 tons, left its imprint in the desert of northeast Arizona near Diablo Canyon. Called Meteor Crater, the crater formed by the impact is nearly 4,000 feet in diameter and is surrounded by a rim that stands 140 feet above the surface. The bottom of the crater is over 500 feet below the rim. Geological studies show that this monster rock struck the Earth 50,000 years ago.

Meteorites striking the Earth may also have a profound effect on Earth life. A theory proposed by planetologists Luis and Walter Alvarez suggests that a massive comet or meteorite (or group of comets or meteorites) crashed into the Earth some 65 million years ago, causing tons of dust and debris to fly into the atmosphere and block out the Sun. The resulting change in climate caused the mass extinction of various plant and animal species—most notably the dinosaurs. This theory remains controversial. Not only is it difficult to accurately date the fossil remains, but how did such a large swarm of comets or meteorites reach the Earth?

There are presently close to 100 known meteor craters on the Earth and many more geological formations that may have been created by the impacts of meteorites. These include the 35-mile-wide Manicougan crater in Quebec, Canada, and the giant 65-mile-wide crater in Popigai, in the former U.S.S.R. Most likely, the Earth has been hit by many more large meteorites since it was first formed, but their craters have been erased by Earth's active surface movements and constant erosion.

The Frequency of Observable Meteors

The frequency of observable meteors varies with the time of day, the season of the year, and primarily, whether the Earth collides with some stray meteoroids or with a stream or swarm of meteoroids. Frequency is greatest in the hours after midnight. On the average, twice as many meteors can be seen in the hours between midnight and sunrise than in a similar interval before midnight. During the former period, the observer is on the front side of the Earth as it moves along its orbit, and sees both meteors that are overtaken by the Earth and those that are met head on (Figure 6.9). There is also a seasonal variation. Because of the inclination of the Earth's equator to its orbit, the frequency of meteors is greatest, for observers in northern latitudes, in the fall. An enormous increase in the number of meteors

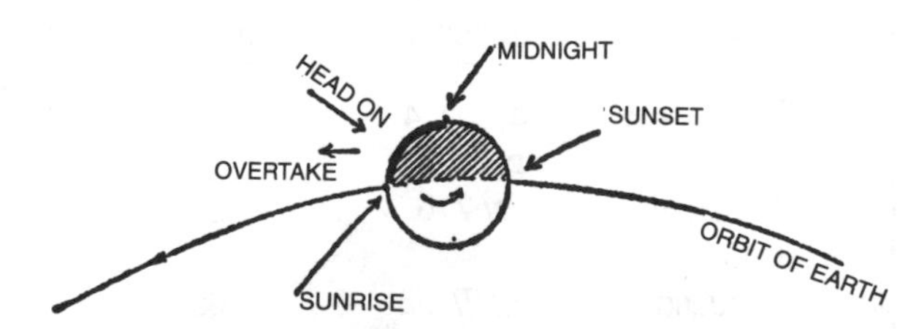

Fig. 6.9. Frequency of meteors after midnight. Nearly twice as many meteors can be seen in any region of the sky after midnight as compared to before midnight. Between midnight and sunrise a terrestrial observer is riding on the front part of the Earth and can see both head on and overtaken collisions between meteors and the Earth's atmosphere.

occurs when the Earth goes through a swarm or a stream, at which time their number may be in the thousands per hour in any small region, as compared to the few that are normally seen per hour. A large number of visible meteors is called a **meteor shower.**

Naturally, meteor showers are much more spectacular when the Earth goes through a swarm than when the Earth goes through a stream, in which the particles are distributed throughout the orbit. On the other hand, passages through streams are much more frequent than passages through swarms. The former occurs annually when the Earth, moving in its own orbit, crosses the orbit of the stream. For a meteor shower due to a swarm to occur, both the Earth and the swarm must be at the point of intersection at the same time. For some swarms, this occurs once in 33 years; for others, it varies.

The meteors in a shower move in parallel paths. Because of perspective, these paths appear to the observer to come together at a point on the celestial sphere. This point is known as the **radiant.** Each shower is named after the constellation in which its particular radiant is located—such as the Lyrids in Lyra, the Perseids in Perseus, and so on. The meteor shower that occurs around the middle of October is created by the particles once in the tail of Comet Halley; while the prominent meteor shower of August is associated with Comet 1862 III. Several of the principal meteor showers are listed in Table 6.1.

TABLE 6.1 METEOR SHOWERS

Name	Date of Maximum	Number per Hour at Maximum
Quadrantids	January 4	110
Lyrids	April 22	12
Eta Aquarids	May 2–6	20
Delta Aquarids	July 27–28	35
Perseids	August 12	68
Orionids	October 21	30
Taurids	November 8	12
Leonids	November 17	10
Geminids	December 14	58

Types of Meteorites

Recoverable meteorites can be grouped into three distinct classes according to their composition:

1. *Siderites, or iron meteorites* These are made up of iron, and 5 to 15 percent nickel, with the two metals usually forming an **alloy,** or mixture. Quite often, a small percentage of cobalt or tiny quantities of other elements are found in siderites. This class of meteorite is the easiest to identify. Siderites usually have a pitted, brownish appearance when discovered, due to the melting the metal underwent while in the air, and to normal rusting on the ground.

2. *Aerolites, or stony meteorites* These largely resemble terrestrial stones, although they are usually denser than the Earthly variety. If a small piece of a suspected aerolite is crushed, glistening flakes of metallic material (nickel, in particular) indicate it probably came from outer space. If small round particles called "chondri" are seen under a microscope, this would confirm the meteorite's extra-

terrestrial origin. There is a sub-group of meteorites called "carbonaceous chondrites" that are made up of original material dating from the formation of the Solar System. This type of meteorite also contains certain organic compounds that are the building blocks of life. (Although they do not contain any life-forms.)

3. *Siderolites, or stony iron meteorites* These meteorites usually are made up of iron-nickel sponge-like frames containing stony material in the connecting structure. The metals and the stones are mixed in about equal proportion. A chemical analysis reveals that they are abundant in iron, oxygen, magnesium, aluminum, and silicon. This mix is quite similar to that in the Earth's crust.

Scientific Interest in Meteorites

Before the Apollo landings on the Moon, meteorites provided the only samples of extraterrestrial matter available to science. Studies of them have provided scientists with two very important pieces of information concerning the origin of the solar system. (See Chapter 8.)

Using the tool of radioactive dating, the age of meteorites has been estimated to be 4.6 billion years. Since these are the oldest known objects, their age is thought to be the lower limit of the Solar System's age—and therefore the Earth. This agument was strengthened when several lunar rocks brought back by the Apollo astronauts were found to be the same age. Scientists have also recovered several meteorites from Antarctica that they suspect may have originated on Mars. The composition of meteorites has also given scientists a better picture about the original state of the materials in the solar nebula. Since meteorites are essentially inert as they travel through the vacuum of space, they have not been changed by air and water erosion or a long history of other chemical changes.

CHAPTER SEVEN

The Nearest Star: Our Sun

KEY TERMS FOR THIS CHAPTER

absolute temperature (same as Kelvin scale)
aurora (pl., aurorae)
Doppler Effect
galaxy
G2
helioseismology
photons
polarity
thermonuclear fusion

Compared to many of the billions of stars in our **galaxy,** the Sun is an average star. It accounts for 98 percent of the total mass of the Solar System and its apparent brightness and size are due entirely to its closeness to the Earth. The next nearest star, α (alpha) Centauri, is almost 300,000 times farther from the Earth than the Sun. It is the third brightest star in the sky and, by sheer coincidence, is almost identical to our Sun in its make-up.

The Sun was formed some 4.6 billion years ago, is nearly halfway through its life, and will not change noticeably for 5 or 6 billion years when it becomes a red giant (see Chapter 8). It is an ordinary celestial body similar to the countless other stars classified as **G2,** or yellow dwarfs. But as far as we know it has one unique feature: It provides our planet with all the necessary energy for life. In addition, it allows astronomers to study at close range some of the complex dynamical processes powering the billions of other distant stars. Thus, the Sun is the indirect source of a large body of astronomical knowledge.

Of the vast amounts of data concerning the Sun, two facts seem particularly startling. One is that the Sun, unlike the Earth, is completely gaseous. A boundary called the photosphere (sphere of light) exists between the Sun and its atmosphere. This is where sunlight originates. The photosphere is quite opaque, making it impossible to see anything beneath it. The solar atmosphere is made up of three layers: The reversing layer, the chromosphere, and the corona, all fairly transparent to the light given off by the photosphere (Figure 7.1). The other startling fact is that the Sun does not spin on its axis at constant angular speed. A point on its equator completes a full revolution in 25 days; while a point 60° north or south of the equator takes 31 days to complete a revolution.

Distance and Diameter

The average distance of the Sun from the Earth is 93 million miles. It is less in January than in July by as much as 3 million miles. The diameter of the Sun is 864,400 miles. The solar diameter is more than a hundred times larger than that of the Earth. The Earth, together with the Moon revolving around it in its orbit, could easily be contained by the Sun.

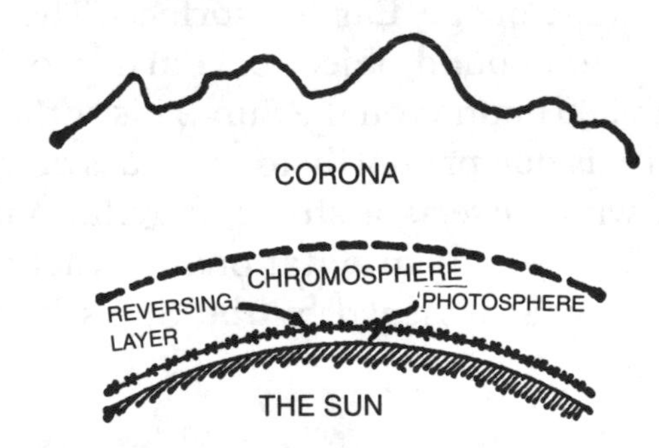

Fig. 7.1. The solar atmosphere. Three layers are recognized in the solar atmosphere; no sharp and definite boundaries exist between them.

The layer immediately above the photosphere (the surface of the Sun) is known as the reversing layer. It is only 1,000 miles thick, but affects the quality of light given off by the photosphere by removing some components from it.

The intermediate layer about 6,000 miles thick is known as the chromosphere. It is here that prominences and flares originate.

The outer layer is known as the corona. It forms a pearly gray layer a million or so miles thick around the Sun.

Compared to other stars, the Sun is of average size. Very small stars have diameters of only 4,000 miles; while very large stars have diameters estimated to be nearly 3,000 times greater than that of the Sun.

Volume and Mass

Given the diameter of the Sun, its volume can be determined. It is 1,250,000 times greater than that of the Earth. The mass of the Sun has been computed as 2×10^{33} grams, or 2×10^{30} kilograms, or approximately 4.5×10^{30} pounds, or 2.2×10^{27} tons, or more than 2 billion billion billions of tons. That is one third of a million times more massive than the Earth. Placed on a balance, the Sun would equal 333,400 Earths in weight.

Compared to other stars, however, the Sun's mass is rather average. There are stars close to a hundred times as massive; others have a mass $\frac{1}{25}$ or less that of the Sun. The vast majority of stars, though, have masses within the range of five times to $\frac{1}{10}$ the mass of the Sun.

Rotation of the Sun

Sunspots also yield information about the spinning of the Sun on its axis. The spinning of the Sun on its axis is based on two facts: All sunspots move across the Sun in the same direction, and sunspots are behind the solar disk for as much time as they are in front of it.

Chief among other proofs is based on the **Doppler Effect** in which the motion of an object changes the frequency of electromagnetic radiation that it emits relative to the observer. The spectra of light from opposite sides of the Sun show marked differences. Light from the edge of the Sun is going away from the observer, indicating a red shift; light from the limb of the Sun approaching the observer indicates a shift of lines toward the blue end of the spectrum.

The direction of the Sun's spinning is the same as that of the Earth. For an observer on the Sun, the stars would rise on the eastern horizon and set in the west. Or, an observer outside the Sun, looking down at its north pole, would see it spinning in a counterclockwise direction.

The period of one complete rotation around its axis is not the same but varies with the latitude. At the equator, it is roughly 26 days; 28 days at a latitude of 45° and still longer at higher latitudes. There is strong evidence that rotation rates also vary at different depths within the Sun. Astronomers believe that the Sun's varying rates of rotation play a key role in the creation of the strong magnetic fields found within sunspots.

Furthermore, sunspots yield information about the tilt of the Sun's axis, based on the slight curvature in the route followed by the spots. The solar axis is inclined by a little more than 7° to the ecliptic. In March, the Sun's north pole is tilted away from the Earth, in September, toward the Earth.

Density and Surface Gravity

The density of the Sun (or any other body) can be found by dividing its mass by its volume. The result of such a computation can be stated in two ways: The density of the Sun is 1.4 times that of water. A cubic foot of its matter would weigh 87.4 pounds. Or the density of the Sun is about .25 that of the Earth.

The density of the Sun is not constant. The density closer to the center is much greater than near the surface because of the sheer weight of the overlying matter. The weight of the material of which the Sun is made would cause a pressure of about 10 billion atmospheres at the center of the Sun, and the density of the gas at the center is 100 times that of water, or nearly 10 times that of lead. Despite this pressure and density, the Sun is gaseous throughout because of the high temperatures at its center—reaching about 14 million degrees **absolute. (Absolute temperature** is the same as the Kelvin scale.) To compare, the absolute temperature of a pleasant day on Earth is 300° absolute.

The surface gravity on the Sun is 28 times that of the Earth's surface. Therefore, a ten-pound object on Earth would weigh 280 pounds on the Sun. This surface gravity is due primarily to the mass of the Sun, which exerts a strong gravitational pull. This strong gravitational attraction is somewhat lessened by the Sun's large radius.

The Interior of the Sun

At the very heart of the Sun is the nuclear reactor. Every second, a process known as **thermonuclear fusion** converts about 4 million tons of hydrogen into helium. This process produces the heat and light so necessary for life on the Earth. Temperatures in the core are estimated to be about 15 million degrees absolute. The energy created within the core is transported toward the surface by a process known as radiation, or the flow of **photons.** This area within the Sun's interior is called the radiative region.

From a depth of about 90,000 miles and then outward to the surface, the Sun's energy moves by a process known as convection. This process is primarily the continuous circulation of hot and cool gases. That is, hot gases rise from the interior, release their energy, are thus cooled and begin to sink back toward the interior. There they are reheated, and then begin to rise again to repeat the cycle. This area is known as the *convective region*. (Watch the circulating bubbles in a pot of boiling water to visualize how convection works.) On the surface of the Sun, these areas of rising and sinking gases look like the cells in a beehive and have been called granulations (Figure 7.2).

The Photosphere

Looking at the Sun through the proper solar filters, one sees a bright disk known as the photosphere—the surface of the Sun.

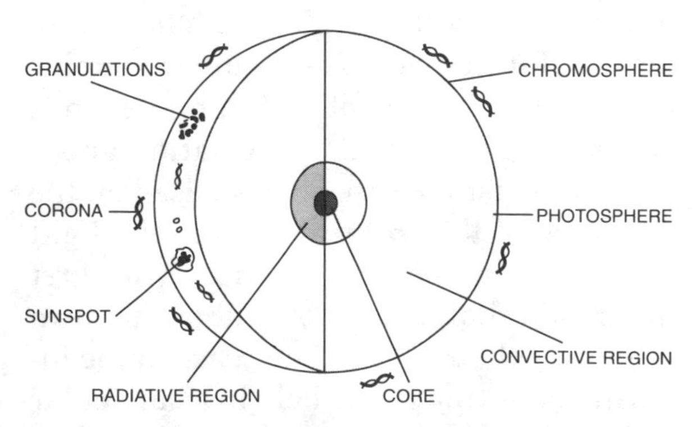

Fig. 7.2. The structure of the Sun. A cutaway view of the internal structure of the Sun.

Actually, the photosphere is not a surface, but rather a shell about 150 miles thick from which light is emitted into space. Light originating beneath this layer is totally absorbed within the 150-mile thickness; little light is produced in the rarefied gases above the photosphere. The top of the photosphere has an average temperature of less than 6,000° Kelvin (abbreviated as K) and a density of .001 of the density on Earth of air at sea level.

The uniform brightness of the disk is only approximately correct. Careful study of the photosphere reveals that this layer is not uniformily bright but rather speckled or marked by granulations, with diameters hundreds of miles long. These granules, probably covering the whole area of the photosphere, are not fixed on the surface; they constantly change in size and structure.

The heat energy produced in the core of the Sun is carried to the surface in part by radiation, in part by convection (or by moving gases or by upward-moving hot bubbles and downward-moving cool ones). This up and down motion breaking through the surface is responsible for the granulations.

Also appearing from time to time are sunspots and faculae. Sunspots are gigantic areas on the solar disk that appear dark by comparison to surrounding regions and have diameters of tens of thousands of miles. Faculae are areas on the surface of the Sun that appear brighter by comparison to surrounding regions.

Sunspots

First referred to by the early Greek observer Theophrastus in the fourth century

B.C., mention of sunspots did not reappear until 1611 when Galileo and others studied them with the newly invented telescope. It was the telescope that shattered the long-held belief that the Sun was a perfect, unblemished sphere. When photographed under excellent seeing conditions, sunspots look like irregularly shaped holes or craters in the Sun's surface. Today we know a great deal about sunspots.

Structure. Most spots consist of two portions that differ greatly in darkness. The inside portion—the technical name is *umbra* (Latin for "shadow")—is the darker of the two. Surrounding the umbra is the semidark portion, the penumbra. "Dark" and "semidark" are relative terms here. Actually, the dark umbra gives off a brighter light than even the most powerful electric light. The area appears dark against the background of the still brighter solar disk. If the sunspot could be detached from the solar disk and viewed separately, it would appear as an intensely bright object. The umbra is 2,000° K, cooler than the rest of the photosphere, but its temperature is still a tremendous 4,000° K (Figure 7.3).

Size. Sunspots vary greatly in size—from 2,000 miles across to more than 10 times that figure. The average size of sunspots is about 6,000 miles, but there have been spots as large as 90,000 miles. Sunspots usually occur in pairs or in complex groups.

Latitude. Spots occur in two belts of the solar surface: One, between 5° N and 40° N solar latitudes; the other, between 5° S and 40° S solar latitudes. There are, of course, exceptions to this rule.

Duration. Small sunspots last for a few days or even a week, and the largest sunspots may last for many weeks. These spots last long enough to be carried across the entire face of the Sun during its rotation and reappear on the Sun's opposite limb.

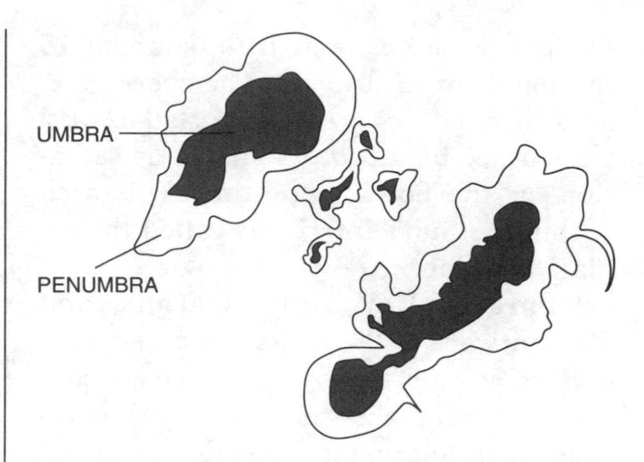

Fig. 7.3. A sunspot group. Sunspots often occur in pairs—one magnetic north and one magnetic south.

Magnetic Field. Each spot is the center of a magnetic field, the strength of the field varying with the size of the spot. Some spots have a "north-seeking" **polarity,** or magnetic alignment; others, the opposite polarity. Studies of magnetic fields are based on the Zeeman effect. (In 1896, Pieter Zeeman [1865–1943] of Holland, discovered the effect of a magnetic field on spectral lines.) Simply stated, spectral lines either split into several components or widen considerably under the influence of a powerful magnet. The manner of splitting or the amount of widening depends on the magnetic field. Information about the magnetism of sunspots is based on the widening of the spectral lines in the light coming from the spots. Indeed, the first indication that a sunspot is about to form is a many-thousandfold increase in the intensity of the magnetic field in a particular area. Also, as the sunspots grow, so does the intensity of the magnetic field. This magnetic field outlives the sunspot by many days, weeks, or months.

Shape and Movements. To the best of our knowledge, the sunspot has the shape of a vortex, its motion being counterclockwise in the northern hemisphere of the Sun and clockwise in the southern. Gases are flowing outward at the base of the vortex and inward at upper levels. The height of the vortex may be 100 miles and magnetic effects are most likely the main driving forces of the gases.

Origin of Sunspots. If sunspots are cooler regions on the surface of the Sun, then why doesn't the heat from surrounding areas leak into them and eventually erase them? The answer seems to be connected with the very strong magnetic fields that are at the heart of every sunspot. All the particles in the Sun's interior are electrically charged. These particles carry heat from the Sun's interior to the surface in the form of convection currents. The charged particles within the sunspots are prevented from escaping and are bent in a circle, just as the charged particles outside the sunspots are prevented from entering by the strong magnetic fields. This explains why a sunspot is cooler when first formed. But, it is still unclear to astronomers why the sunspot remains cooler when heat—in the form of uncharged photons, unaffected by the magnetic field—should enter the sunspots and warm them up in a few days. A complete theory explaining sunspots still remains a major challenge for astronomy.

The Sunspot Cycle

For more than 200 years, observers have recorded the number of sunspots appearing on the face of the Sun. As early as 1843, a theory of sunspot cycles was suggested and has since been confirmed. The average period of a complete cycle is 11 years; that is, the time between the maximum or minimum number of sunspots is about 11 years, although intervals as short as 7, or as long as 17 years have been recorded. A period of peak activity, or solar maximum, occurred in 1991 to 1992.

There seems to be evidence, first proposed in the 1890s by E.W. Maunder, an English astronomer, that the Sun undergoes a variation in solar activity over a much longer period of time—as much as several hundred years. These Maunder cycles seem to have some relationship with climatic conditions on the Earth. For example, between 1645 and 1715, during the so-called Maunder minimum, the Earth experienced an unusual cooling of its atmosphere.

Flares and Prominences

Some of the most spectacular sunspot-related features are flares, prominences or filaments. Flares are tremendous and explosive outbursts of energy on the surface of the Sun that hurl particles and radiation into the Solar System. Within a few minutes, they release the same energy as a billion-megaton bomb and are some of the most violent disturbances on the Sun. Most frequent during peak sunspot activity, large solar flares may interfere with radio communication on the Earth. Flares also often produce changes in the Earth's own magnetic field. They are also responsible for the colorful draperies of our **aurorae** (the Aurora Borealis, or northern lights, and Aurora Australis, or southern lights) most frequently seen above 70° north or 70° south latitudes. An aurora is produced when electrically charged particles emitted by the Sun during periods of flare activity strike the Earth's upper atmosphere causing atmospheric gases to glow.

As masses of bright gases that appear in the corona far above the Sun's surface, prominences are cooler and denser than the surrounding corona and are reasonably bright. Sometimes prominences rise upward in great surges and eruptions; sometimes they seem to stream downward from great heights. Often the glowing streams of gas form graceful arcs that are formed by lines of magnetic force looping upward from the chromosphere. Many prominences are stable enough to seem to float for hours or days above the solar surface. Prominences also are called filaments. Such prominences appear as dark, snakelike filaments.

Faculae or Plages

Plages (French for "beaches"), or faculae (Latin for "small torches"), emit light across the entire visible spectrum. They are bright patches found directly above sunspot groups. A plage usually precedes the appearance of a sunspot in a high-magnetic field on the Sun. It remains while the sunspots are visible and lasts for several weeks or more after the sunspots disappear. Current belief is that the gases in a plage have been heated to a very high temperature by the concentrating energy of a strong magnetic field in that region of the Sun.

The Layers of the Sun

The Reversing Layer

This is the lowest of the three layers in the Sun's atmosphere. The base of the layer is the surface of the Sun. The top extends to about 1,000 miles above the surface. This thickness has been determined from studies of solar eclipses, the time it takes the Moon to cross that layer plus the known value of the Moon's speed.

The reversing layer is responsible for the many thousands of dark lines in the otherwise continuous spectrum of sunlight (see Chapter 2). This is because the gases in the reversing layer absorb these particular wavelengths. Among the 65 elements present in this layer are hydrogen, carbon, nitrogen, oxygen, aluminum, iron, cobalt, cadmium, lead, and platinum.

The Chromosphere

The name for the middle layer in the Sun's atmosphere is the chromosphere, with an average thickness of about 6,000 miles. In some zones of the Sun the thickness may be as much as 8,000 miles; in others, as little as 5,000 miles. The chromosphere, or so-called color sphere, owes its name to its very bright color (rose-pink). This is largely due to a line in the spectrum of hydrogen, labelled "H-alpha" (6,563 angstroms). This also is the layer in which most of the previously described solar flare and prominence turbulence occurs.

The Corona

The corona is the uppermost layer of the solar atmosphere, only visible to the naked eye during a total eclipse of the Sun. It can also be seen and photographed at other times with special telescopic equipment such as a coronagraph, which produces an artificial solar eclipse. The corona is a pearly-gray halo of complicated design surrounding the body of the Sun. It is considerably larger than the two layers beneath it, being close to a million miles in thickness.

The shape of the corona is closely connected with the 11-year period of sunspot activity. At sunspot maximum, it is circular and has few pronounced rays. At sunspot minimum, it is elongated and enormous streamers appear to radiate from it (Figure 7.4). This is primarily due to the magnetic fields associated with the formation of sunspots. At the solar poles, the corona almost disappears because the Sun's magnetic fields are weakest in those regions. Therefore, solar particles (electrons, protons, and heavier ions) are able to escape into space as the so-called solar wind through holes in the corona. These holes have been detected and photographed in the X-ray spectrum by such space missions as Skylab in the 1970s and the Solar Maximum Mission of the 1980s.

Fig. 7.4. The corona. A drawing of the corona during the eclipse of the Sun in 1900. Sunspots at the time were at a minimum and the corona was fairly elongated. The thickness of the corona is greatest in the equatorial regions of the Sun.

One of the remarkable characteristics of the corona is that it has an enormously high temperature. The temperature, arrived at by several independent methods, is in the neighborhood of several million degrees Kelvin. Research points to the continuous shocking and compressing of the coronal gases by turbulences rising out of the boiling surface of the Sun as a possible cause of such high temperatures. Another theory states that the intense magnetic fields on the surface heat the coronal gases to such temperatures.

The Spectroheliograph

Much of the knowledge about the Sun and its atmosphere was obtained with the aid of an instrument known as a spectroheliograph. Introduced by Professor George E. Hale (1868–1938) in 1890, it has been invaluable, allowing astronomers to obtain the distribution of any element on the disk of the Sun. In a few moments an astronomer can determine the distribution of hydrogen, oxygen, calcium, or any other element on the part of the solar surface facing the Earth.

The spectroheliograph determines not only the location of the element on the

Sun's surface but also the nature of its motion. Spectroheliograms taken of sunspot areas indicate, for example, the whirling motion of hydrogen gas.

This instrument is used primarily with an image of the solar disk. It cannot be used with stars, as they appear only as points of light—even at great magnification. It is of little use with planets because the light from the disks is reflected.

Current Solar Research

Even though the Sun has been studied many years, astronomers today still do not fully understand the internal workings of our closest star. The discovery that the solar surface constantly quakes led astronomers to a way to probe the interior of the Sun. By studying the seismic waves that shake the surface, astronomers are forming a new picture of the Sun that may answer many of its mysteries—and those of stars in general.

There are many questions. How hot is the Sun's core? Does the core rotate and how fast? How deep is the convective region? Where are the majority of the solar neutrinos—tiny, massless particles generated in the Sun's thermonuclear reactions? What are the densities of each layer from the center out? Or is the diameter of the Sun contracting or has it remained the same over the past several billion years? These questions are the prime targets for this new science of **helioseismology.**

Within the last 10 years scientists have been trying to see beneath the solar surface by carefully studying how the sound waves created by solar quakes travel through the Sun's interior. This is essentially the same method that geologists use to study the Earth's interior—studying seismic waves created by terrestrial earthquakes. These studies have shown that the entire Sun is ringing with sound waves like a huge bell. The Global Oscillation Network Group (GONG) is a worldwide project directed from the National Solar Observatory in Tucson, Arizona, that is planning to coordinate and study helioseismological data. The group hopes to be in full operation by the mid-1990s and obtain a more refined picture of the Sun.

The Ulysses Mission

On October 6, 1990, the Ulysses space probe became the fastest human-made object ever sent into space. Its five-year mis-

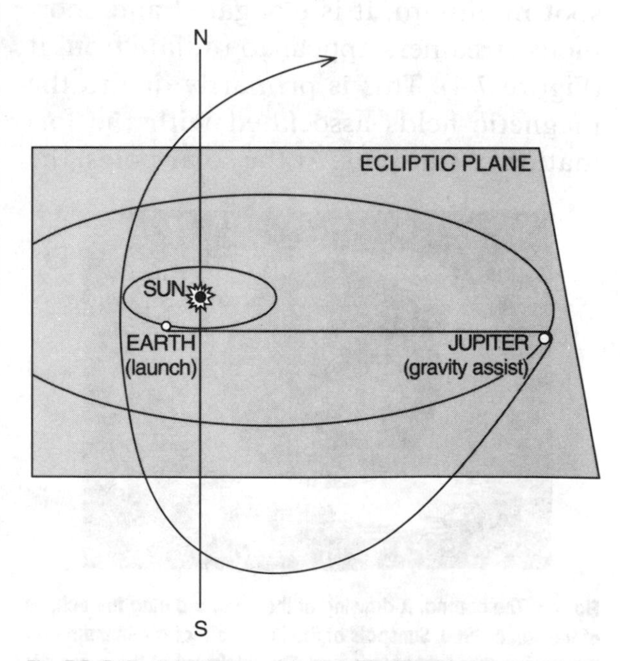

Fig. 7.5. Path of the Ulysses mission. The Ulysses spacecraft was designed to explore the regions around the North and South Poles of the Sun.

sion includes a flyby of the planet Jupiter where it will receive a much-needed gravity boost in speed in order to accomplish its primary goal—the exploration of the northern and southern polar regions of the Sun.

Ulysses carries five European and four American instruments designed to investigate the following:

- Properties of the solar wind
- The area where the solar wind rises from the surface of the Sun
- The solar magnetic fields
- Solar radio bursts and plasma waves
- Solar X-rays
- Solar and galactic cosmic rays
- Gas and dust in interplanetary space

As Ulysses passed 260,000 miles from Jupiter's cloudtops in February 1992, the planet's gravity shot the probe into uncharted territory, the region of the Solar System below the plane of all the planets' orbits. (Because of the tremendous energies needed to overcome the gravity of the Sun, all previous space missions have been confined to the narrow plane where the planets orbit the Sun.) The spacecraft will reach 80° south solar latitude in June 1994 and begin its flight under the Sun's south polar region at a safe distance of 2 AU from the Sun's surface. Ulysses will then head north, crossing the Sun's equator in February 1995, followed by a four-month pass over the Sun's north polar region beginning in June 1995. The mission is scheduled to end in September 1995 (Figure 7.5).

CHAPTER EIGHT

The Origin of the Solar System

KEY TERMS FOR THIS CHAPTER

accretion	*radius*
angular momentum	*red giant*
centrifugal force	*rotational velocity*
proto-	*white dwarf*

It is generally believed that our Solar System began just over 4.6 billion years ago when a huge cloud of gas and dust, called the solar nebula, began to collapse into a circular, spinning disk. At the center of this cloud was a concentrated knot of material known as the **proto**-Sun. Later in its development, this proto-Sun would ignite to become our Sun. As the nebula condensed, pockets of material began to stick to each other, forming hundreds of bodies called planetesimals, or small planets. The planetesimals grew as the **accretion** process continued. The planetesimals in the Inner Solar System condensed into bodies of silicates and iron because of the Sun's heat; while in the colder, Outer Solar System, ammonia, water, and methane condensed as solids. These solids grew, or accreted, with dust and other lighter elements to form the gaseous atmospheres around the giant planets.

There have been many theories that try to explain the exact mechanisms behind the formation of the planets, satellites, asteroids, and comets. Chief among these are the centrifugal-force hypothesis, the tidal hypothesis, the collision hypothesis, the double-star encounter hypothesis, the turbulence hypothesis, and the proto-planet hypothesis. This chapter will examine some of the strengths and weaknesses of these theories, while also inviting some of the most recent thinking about planetary formation into the debate.

The Centrifugal-Force Hypothesis

This hypothesis is usually referred to as the nebular theory of Laplace, after the great French mathematician and astronomer Pierre Simon de Laplace (1749–1827), who first proposed it in 1796. According to this theory, the Sun was once a vast, slowly rotating, disk-shaped mass of hot gas, extending well beyond the present orbit of Pluto. The birth of the planets resulted from the following chain of events:

1. The gas cooled.
2. The gas contracted, with the **radius** of the disk becoming smaller.
3. The decrease in radius caused an increase in **rotational velocity,** or rate of spinning, thus increasing the **centrifugal force.**
4. When the centrifugal force, acting on the outermost regions of the Sun, grew greater than the force of attraction, a ring was separated from the main body of the Sun.
5. The gaseous ring gradually condensed into a sphere which became one of the planets.
6. The Sun continued to cool, and this process repeated itself to produce the other planets (Figure 8.1).

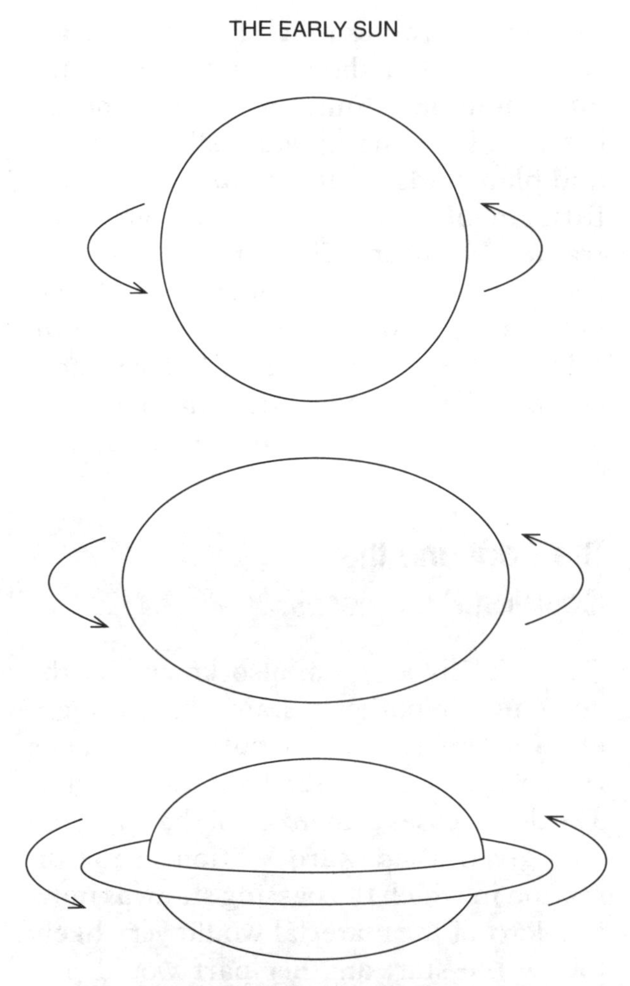

Fig. 8.1. The centrifugal-force hypothesis. According to this theory, the Sun was once a vast rotating disk of hot gas that extended beyond the present orbit of Pluto.

At first glance, the theory seems reasonable, but further thought shows it to be completely at odds with several basic principles of celestial mechanics. First, it can be shown that the rings, after being separated from the Sun, could not coalesce to form a single body, nor even a few bodies. Instead, physical theory shows that most of the material in these rings would evaporate, molecule by molecule, into space; the remainder, due to tidal effects from the Sun on the ring, would become countless gravel-size or smaller particles.

Second, the theory suggests that planets so created should rotate and revolve much more slowly than they actually do (or the

Sun should rotate much faster). In other words, the Sun should have most of the **angular momentum,** or rotational energy, of the Solar System; while all the planets and planetoids together should have very little angular momentum. The facts contradict this theory. The Sun, with 99.9 percent of the mass of the Solar System, has only 2 percent of the angular momentum. All the other members of the Solar System combined have .1 percent of the mass, but 98 percent of the angular momentum.

The Tidal and the Collision Hypotheses

The tidal hypothesis, also known as the encounter hypothesis, states that the planets were created as a result of enormous tides raised on the Sun by a passing star. The dense gases drawn out of the Sun were also given a sideward motion in the direction in which the passing star was moving. Part of the material would very likely follow the star; another part would presumably return to the surface of the Sun. The last part would be affected by a centrifugal force large enough to overcome these gravitational attractions, thus forming the planets (Figure 8.2).

This hypothesis, suggested by Forrest R. Moulton and Thomas C. Chamberlin of the University of Chicago in 1900, was originally known as the planetesimal theory. This diminutive suggests that the immediate results of the tidal action was merely the formation of small planets. These grew in size by picking up loose material to form the nine known planets.

The collision hypothesis differs from the tidal hypothesis in that it assumes that the

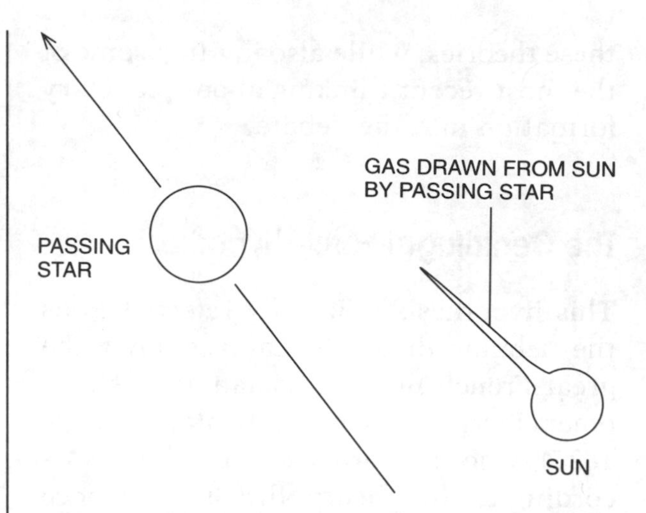

Fig. 8.2. The tidal hypothesis. According to this theory, a passing star drew gases from the Sun that eventually formed the planets.

encounter between the Sun and the visiting star took the form of an actual collision. Again, the problem with these two theories has to do with the distribution of angular momentum; namely, that the Sun has only 2 percent, while the planets have close to 98 percent of this rotational energy. Professor Henry Norris Russell of Princeton University concluded that such an encounter could only produce as much as 10 percent of the known angular momentum per ton of planet. Also the likelihood of a collision between two stars is extremely rare. Computations indicate that no more than ten collisions have taken place among the 200 billion stars in our galaxy in the last 5 billion years.

The Double-Star Encounter Hypothesis

This theory, introduced by the English astronomer R.A. Lyttleton, has the enormous advantage over the preceding ones because it does not go against the observed facts of angular momentum. Lyttleton's hypothesis assumes that the Sun was originally a double star (there are many in the galaxy), and that a passing star collided with the Sun's companion. The events that followed this collision may have lasted about an hour. The colliding stars (the intruder and the companion of the Sun), after the rebound, drew out a ribbon of material, large enough to produce all the bodies in the Solar System (except the Sun). The two stars (like billiard balls after a collision) went along their courses, carrying away parts of their ribbon of gas and dust that were under the immediate influence of their respective gravitational fields. The various members of the Solar System developed from the central part of that withdrawn matter. The angular momentum of the matter in this central part could conceivably correspond to the observed values.

Computations show that 94 percent of the ribbon would be dragged along by the two colliding stars; while the 6 percent of the central portion would follow the intruder and the Sun's companion for some time before returning to the two stars. The likelihood of the planets forming from this 6 percent of the matter is extremely slight. It is infinitely more likely that tidal effects from the intruder and the companion tore the ribbon to shreds and eventually scattered it into neighboring space.

The Turbulence Hypothesis

One of the most promising hypotheses is the work of the twentieth-century German physicist Carl Friedrich von Weizsäcker. Published in 1945, his theory is similar to that of Laplace. Von Weizsäcker assumes that at one time in its development, the Sun was surrounded by a slowly rotating disk-shaped cloud of gas. The diameter of this disk was equivalent to the diameter of the present solar system, and the temperature at various distances from the central Sun corresponded to the temperatures that now exist on the nine planets located at the same distances.

The mass of the nebula is assumed to have been 100 times larger than the combined masses of the planets, equal approximately to 10 percent of the Sun's mass. The nebula was made up of primarily hydrogen and helium (99 percent) and only 1 percent of the heavier elements. Weizsäcker gives the following chain of events: Over the course of 200 million years, the molecules of hydrogen and helium thinned into neighboring space, decreasing the mass of the gases from the original 10 percent to the present value of slightly more than .1 percent of the Sun's mass. The angular momentum of the nebula was basically unchanged, thus accounting for its present large value. In those 200 million years, due to the differences in speed between parts of the nebula close to the Sun (great speeds) and those farther from it (less speeds), turbulence cells were formed. The matter in each cell moved in a clockwise direction, while the cells themselves moved counterclockwise (Figure 8.3). The planets, according to this hypothesis, formed in the dead-end region between the cells. These regions, due to the conflicting currents at their boundaries, were most active in accretion of material from the neighboring cells, and were thus instrumental in building up great amounts of matter rotating in a counterclockwise direction. Five such equidistant quantities

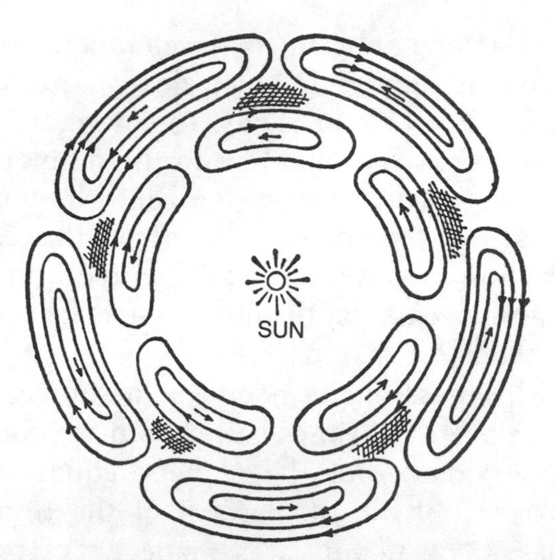

Fig. 8.3. The turbulence theory. According to this theory, differences in speed in the various parts of the solar nebula formed turbulence cells. Large masses of matter formed in the regions between the cells (shaded areas in the diagram). These lumps eventually formed planets.

eventually joined in the formation of the planet at the given distance.

Much that we know about the Solar System can be explained by Weizsäcker's hypothesis. The hypothesis explains, for example, the fact that the planets, as well as the planetoids, revolve in nearly a common plane; that the Sun's equatorial plane nearly coincides with the orbital plane of the planets; that the planets have nearly all of the angular momentum of the whole system; and that the spacing of the planets follows a regular pattern, as described in the Titius-Bode law for planetary distances.

Many questions, however, still remained unanswered. Among these were how did the droplets and particles form the lumps? By what process did the several lumps unite to form a single planet? Why does Uranus rotate almost perpendicular to the axis of its orbital plane? Why do a few of the satellites move in retrograde motions?

The Proto-Planet Hypothesis

Proposed in 1950 by the Dutch-born American astronomer G.P. Kuiper, the proto-planet hypothesis modifies several parts of the nebular theory of Laplace. According to this hypothesis, the large sphere of gas and dust, under the influence of gravitational and centrifugal forces, became a fast-rotating disk. Ninety-five percent of the original material settled near the center of the disk (this material was soon to become the young, cool Sun), the other 5 percent remained in the disk (soon to become proto-planets of the Solar System).

Turbulence was the rule in the disk, and concentration of matter formed in various parts of it, only to be dissolved soon after formation. At one time or another, though, a concentration of matter did form in which the gravitational attraction was strong enough to overcome the disruptive forces of the turbulence. Such a concentration grew rapidly in size and mass by adding neighboring matter to itself, eventually becoming a proto-planet. The other proto-planets were similarly formed at various distances from the Sun. The proto-planets close to the Sun accumulated little mass, since most of the material at that distance was already picked up by the Sun. Proto-planets far from the Sun did not grow much in size, since they formed at the outskirts of the disk where the amount of material was limited.

The young, cool Sun continued to shrink and warm up. Eventually, about 5 billion years ago, the core became hot enough to trigger the hydrogen into helium fusion, and the Sun began to give off radiation as well as solar wind at full power. The radiation and the solar wind swept out all the material between the proto-planets in the system, and heated the proto-planets, thus causing a great deal of their mass to escape into interstellar space.

In this sweeping-out process the Earth lost 99.9 percent of its mass as a proto-planet (the mass lost was mainly hydrogen and helium). The proto-planet Jupiter, too, lost nearly all its mass (95 percent) in the process of changing into a planet.

Current Thoughts on Planetary Formation

Much of the current thinking about the origin of the Solar System is based on the hypotheses above. Today's astronomers, however, use the latest computer simulations to recreate and study their ideas about planet-shaping events that happened billions of years ago.

According to some specialists, collisions among many of the early Solar System planetesimals were the major instruments in the creation of the planets and satellites as we know them today. For example, the planetesimals in the Inner Solar System ranged from Mars- and Moon-size to much smaller pieces. The larger chunks eventually collided and accreted to form Mercury, Venus, Earth, and Mars.

The planets swept up much of the smaller debris and grew in size. Other pieces of planet-forming materials continued to wander through the Solar System. Some of these chunks remained in orbit in the plane of the ecliptic; others, because of collisions or gravitational tugs from larger planetesimals, were tossed out of the ecliptic into eccentric orbits.

Then, between about 4 billion and 3.8 billion years ago—after the surfaces of the terrestrial planets had cooled enough to become solid—the remaining planetesimals in the Inner Solar System are thought to have heavily bombarded the inner planets. According to an impact-origin theory proposed by H. Jay Melosh at the University of Arizona's Lunar and Planetary Laboratory, Earth's own Moon may have formed in such a manner 3.8 billion years ago. His idea states that a Mars-size planetesimal struck the Earth at low velocity, sending a chunk of the planet into space and eventually forming our Moon.

The collision theory also assumes that the Outer Solar System also experienced an era of heavy bombardment, but its precise time of occurence is not known. The evidence of the bombardment can be seen on many of the satellites in the Outer Solar System. The whole era of bombardment is collectively known as the late heavy bombardment.

The current theories of planetesimal collisions can also explain the formation of the asteroid belt. The planetesimals did not become a planet in the space between the orbits of Mars and Jupiter—although there should have been enough material to form one. The tremendous gravity of Jupiter in this area caused these planetesimals to collide violently rather than to coalesce. A constant grinding of asteroids against one another wore the particles down to smaller sizes and threw some of the fragments out of the belt altogether.

There is strong evidence that a very large chunk of material probably crashed into the planet Uranus and tipped it onto its side. This may explain, also, why several satellites appear to revolve in either highly eccentric orbits or rotate in retrograde.

The collisions of countless planetesimals with the inner planets may have been instrumental in the formation of the planets' early atmospheres. The intense heating that resulted from repeated bombardments could have vaporized and released many of the primitive gases from under the crusts and into the early atmospheres of Venus, the Earth, and Mars. Icy planetesimals from beyond the orbit of Jupiter could have been sent toward the Sun by that giant planet's gravity. Upon crashing into the inner planets, they may have contributed water and other elements to the atmospheres of those planets.

Interestingly, there is a theory proposing that life could be spread by the collisions of planetesimals, especially by icy comets. Called "panspermia," this theory states that micro-organisms such as bacteria could remain dormant and protected from the rigors of space for long periods of time until they find the right conditions to flourish. Most scientists believe that the logistics for such a spreading of life are too difficult and that the chance of finding the right conditions to flourish are unlikely.

The Future of the Solar System

To the best of our knowledge, the major change in our Solar System will be the aging of the Sun. The Sun is now in its maturity (see Chapter 9 for more information on the evolution of stars) and receives its energy from the thermonuclear process—or the reaction that changes hydrogen into helium. This process will probably continue for another several (five or more) billion years.

Next, the Sun will start on its way to becoming a **red-giant** star. At that time, the following events will likely occur: The Sun will grow bigger, possibly swallowing Mercury and even Venus. Its surface will cool and appear redder. It will increase, by a factor of 100, the amount of radiation sent toward the Earth. As a result of this, the Earth's oceans will boil away and its atmosphere will escape into space. As the Earth takes on all the characteristics of a charred cinder, the nearest place for human comfort might be Saturn or even Pluto!

The red-giant period for the Sun is likely to last several hundred million years and will be followed by a change to a **white dwarf.** At this stage the Sun will grow smaller—eventually smaller than the Earth. It will change color to blue or white and will decrease in brightness by a factor of $1/_{10,000}$ of its present value. Finally, it will appear to a hypothetical Earth observer as a point of light—not as a disk.

The fate of the Earth is, of course, tied to the Sun. On Earth temperatures will drop drastically, even approaching absolute zero. Darkness will be the rule 24 hours a day. The stars will still be in the sky (among them one particularly bright one—our Sun). The planets will be invisible, a very pale Moon will go through its phases, and from time to time a comet will be seen not too far from the particularly bright star.

All of this—the too-hot period and the too-cold period—is billions of years away. Perhaps we can use this time to advance the ethical, moral, and scientific values on the planet now occupied by humanity.

CHAPTER NINE

The Physical Properties of Stars

KEY TERMS FOR THIS CHAPTER

arcsecond
interferometry
occultation
parallax
parsec
proper motion
radial velocity
reciprocal
speckle interferometry
spectroscopic binary
tangential velocity
variable star
visual binary

By the middle of the nineteenth century, new discoveries in astronomy had begun to slow down. The optical telescope had been perfected to a high degree, and no one believed there were other windows in the sky. Many thousands of stars had been catalogued and filled huge astronomical encylopedias. Two new planets were discovered but further research had failed to find any others. Astronomers had learned everything they could about stars as points of light. But they did not know what the stars are made of nor why they shine. The enormous distances of the stars seemed to prevent any further study. Astronomy, the oldest science, was about to enter the graveyard of dead subjects when a German physics professor named Gustave Kirchoff (1824–1887) working with the chemist Robert Bunsen (1811–1899), the inventor of the Bunsen burner, discovered something totally new—flaming objects send out a coded message about themselves!

The discovery was that a hot atom radiates light at a series of wavelengths peculiar to that element. Therefore, the chemical make-up of a star could be determined by analyzing the light coming from the atoms in that star. This is the message of starlight. It shows us what the stars are made of and how they are born, evolve, and die. Kirchoff discovered how to read this code, and with his discovery, the science of astronomy was reborn.

Stellar Surface Temperatures

The spectrum of a star's light is used to determine its surface temperature—the layer known as the photosphere. Stellar temperatures are usually stated using the Kelvin scale, shown by degrees Kelvin or °K. This scale is also called the absolute temperature or °A. 0° K is equal to −273° C, or −459° F. Typical temperatures of stellar surfaces are about 5,000 to 7,000° K. Extremely hot stars, like ζ (zeta) Puppis, have temperatures like 30,000° K, and some stars have surface temperatures as high as 200,000° K. On the other extreme, cool stars like χ (chi) Cygni, a **variable star,** one that periodically changes its output of energy, at the time of its minimum brightness has a temperature of only 1,800° K.

These are the temperatures of the surface layer or photosphere of a star. The temperatures in the interior of the stars are much greater but they cannot be directly measured. Interior temperatures range not in the thousands but in the millions of degrees. These temperatures will be discussed later when we discuss the nuclear reactions that generate energy in the interiors of stars.

Stellar Distances

Some stars are fairly close to us. The light leaving them reaches us in a few years. The tremendous distances of most other stars stagger the imagination, with some stars' light reaching us in millions of years. There are two ways to determine distances of stars: The direct and indirect methods. Direct methods of finding distances to stars rely on stellar objects no further than 300 light years from Earth. Indirect methods rely on special kinds of stars such as Cepheids and RR Lyrae variable stars.

Triangulation, or **parallax,** is one direct method used to find the distance to a star. In this method, often used by surveyors on Earth, a distance is determined by measuring three quantities: the length of an arbitrarily chosen line and two angles. The chosen line is known as the line of position. Using the line of position and the two angles, the unknown distance is computed, using the properties of right triangles (Figure 9.1).

In finding distances to stars, the astronomer is handicapped by the lack of a long enough line of position. The largest line available to the astronomer is the diameter of the Earth's orbit around the Sun (186 million miles), which is only a tiny fraction of the distance to even the closest star.

Using this 186-million-mile line of position, one angle is carefully measured for a nearby star. Six months later, when the Earth is at the opposite point in its orbit

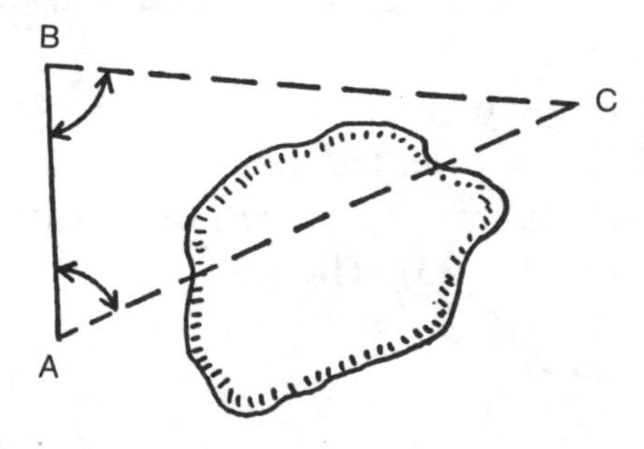

Fig. 9.1. Finding distances by the method of triangulation. To find the distance between points A and C, which may happen to be on opposite sides of a lake, a line of position, AB, is laid out. Knowing the length of the line AB, and the two angles A and B, it is easy to compute the desired length AC.

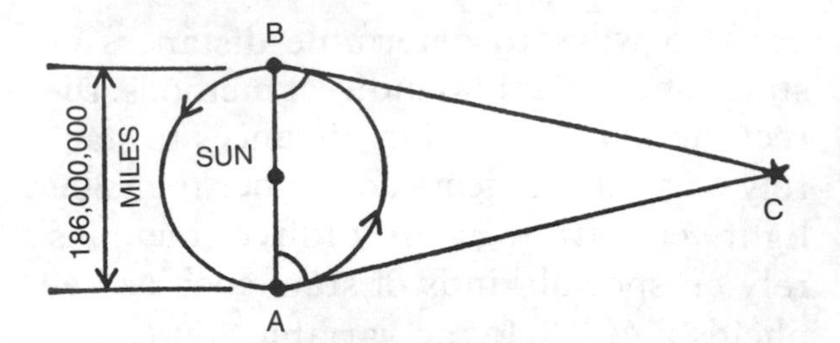

Fig. 9.2. Method of determining the distance from the Sun to a nearby star C.

around the Sun, the other angle is carefully measured. Using this information and proper trigonometric formulas, one can compute the parallactic angle and the distance to the star in question (Figure 9.2). This method, which can only be used to determine the distance to nearby stars, is based on the fact that the more distant stars do not appreciably change their relative positions in a six-month period.

Distance to a star usually means the distance between the center of the star and the center of the Sun. At times the distance between the center of the star and the center of the Earth is used. The difference between the two, the radius of the Earth's orbit, is not significant in measuring stellar distances. By agreement among astronomers, the radius of the Earth's orbit, and not the diameter, is taken as a line of position.

The angle formed by the star on the radius is known as the parallax. The more distant the star is, the smaller is its parallax (Figure 9.3). The parallaxes of stars

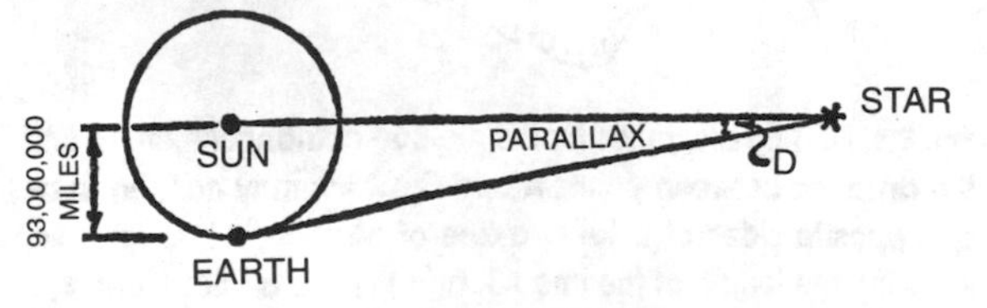

Fig. 9.3. The angle D is known as the parallax.

are extremely small angles. Even the nearest star, α (alpha) Centauri, has a parallax of only 0.756 **arcseconds.** An arcsecond is $\frac{1}{360}$ of a degree. This angle is even smaller than the diameter of a dime would form at a distance of a mile. Other stars form angles of 0.1 arcseconds and less. Measuring such extremely small angles is a very exacting job.

In the process of finding the parallaxes for the various stars many corrections have to be made by the observer. Some of these corrections are now carried out automatically by techniques developed in measuring parallaxes. A variety of these corrections are due to the motion of the star; others are due to the motion of the observer; while still others are due to refraction of light by the Earth's atmosphere.

During the six-month interval between observations, the star itself may have moved slightly, relative to other stars. In the same interval, the whole Solar System, together with the observer, may have changed position. To obtain a reasonable estimate of the magnitude of these corrections, several sets of measurements extending over a period of years are taken for each star under study. From measurements taken a full year apart, estimates can be made on the corrections made necessary by the various motions. Corrections due to the refraction of light by the Earth's atmosphere have to be carefully computed; otherwise, serious errors in the determination of distance may be introduced. Several of these corrections are automatically accounted for, in routine parallax work. The distances to the majority of stars cannot be found by the direct, or parallax, method because the parallaxes of those stars are much too small to be measured even with the best available instruments.

Measurements of Stellar Distances

The distances to stars are so great that the use of ordinary units such as miles is not practicable. The nearest star is 25,000,000,000,000 (25 trillion!) miles away—a number that is rather awkward to write, to remember, and to use. Astronomers instead use a kind of shorthand. Three units of measure are commonly used: the astronomical unit (AU); the light-year (ly); and the **parsec** (pc).

One astronomical unit equals the average distance from the Earth to the Sun—about 93 million miles. This unit is fairly small (astronomically speaking), and is used primarily for describing distances within the Solar System (see Chapter 1). For example, the distance from the Sun to Pluto is 40 AU, or 40 × 93,000,000 miles (3,720,000,000 miles).

A larger unit of astronomical distance is the light-year (ly), defined simply as the distance traveled by a ray of light in a period of one year. (It was already shown in Chapter 1 that one light-year is equal to 5.88 trillion miles.) Using this unit of distance, α (alpha) Centauri is 4.2 ly away. In other words, the light of the star we see has been traveling to the Earth for 4.2 years.

The definition of a larger unit of distance, the parsec, is based on the triangle shown in Figure 9.4. If the parallax D is equal to 1 arcsecond, then the distance between the star and the Sun is 1 parsec. A parsec is an extremely large distance. In terms of miles, one such unit is equal to

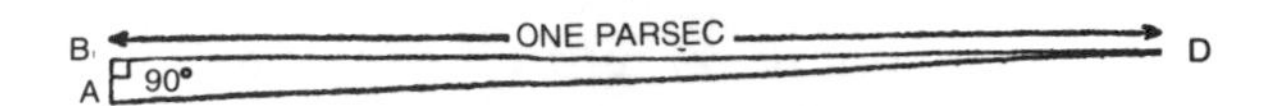

Fig. 9.4. Definition of a parsec. If: (1) The angle B is 90°; (2) the side of AB is 93,000,000 miles long; and (3) the angle D is 1 second, then: the side BD is equal in length to one parsec.

about 20 million million miles. One parsec is 206,265 times as large as an astronomical unit. For example, the closest star, α (alpha) Centauri, is 1.3 parsecs away; while other stars in our galaxy are hundreds and thousands of parsecs away. (It is interesting to note here that when measuring even greater distances beyond our own galaxy, measurements are stated in kilo- ["thousands"] and mega- ["millions"] parsecs.)

The simple relationship between the parallax of a star and its distance in parsecs should be noted. One is given by the **reciprocal,** or inverse value, of the other. For example, a star having a parallax of 0.5 arcseconds is 2 parsecs away; a star having a parallax of 0.2 arcseconds is 5 parsecs away; a star having a parallax of 0.1 arcseconds is 10 parsecs away, and so on. Distances in parsecs can be easily changed into light-year distances. For example, 1 parsec equals 3.26 ly, 2 parsecs equal 6.52 ly, 10 parsecs equal 32.6 ly, and so on. In addition to the Sun and α (alpha) Centauri, the following are among the stars closest to us: Barnard's Star is 6.1 ly away; Lalande 21185 is 8.1 ly away; and Sirius is 8.6 ly away.

Sizes of Stars

Stars come in a huge variety of sizes. The smallest known one (not including the exotic neutron stars that will be discussed in the next chapter), with a diameter of a mere 4,000 miles, is identified by its catalogue number AC+70°8247. The largest known star, ϵ (epsilon) Aurigae B, has a diameter nearly 3,000 times that of the Sun. (Remember that the Sun's diameter is 864,000 miles.)

The diameter of a star cannot be measured directly by using a telescope. Even through a large telescope, stars appear as points of light having no measurable diameter. Scientists have developed methods, however, to measure the true linear diameter of a star. One ingenious method, based on the phenomenon of light waves cancelling each other out, has been used to measure the diameter of the largest, nearest, and brightest stars. Originally suggested by the American physicist Albert Michelson (1852–1931) about a century ago, this method, called **interferometry,** was used as early as 1920 at the Mount Wilson Observatory. Using an interferometer, astronomers can measure the angular diameter of stars, which is the angle formed by the star's diameter as seen by the observer. Once the distance to the star is measured, the distance and the angular diameter can be used to compute the true diameter of the star in arcseconds.

Among the stars that were measured with the interferometer are the variable red-giant Betelgeuse, whose angular diameter varies from .034 to .042 arcseconds; Arcturus, and Aldebaran, each indicating an angular diameter of .020 arcseconds. As the distances to these stars are known, their linear diameters can be computed by multiplying the angular diameter by the distance. The diameter for Betelgeuse is equal to 800 times the diameter of the Sun; while Arcturus and Aldebaran are respectively 25 and 40 times the diameter of the Sun.

The size of a star can also be computed by indirect measurements. One such indirect method is based on the relationship between the luminosity (L), temperature (T), and diameter (D) of a star. Luminosity is the measure of a star's true brightness and is usually stated in some multiple of our Sun's luminosity (the Sun has a luminosity of 1). Among the most luminous stars is S Doradus in the Large Magellanic Cloud (a neighboring galaxy of the Milky Way). S Doradus' luminosity is 600,000 times that of the Sun; that is, if the Sun and S Doradus were placed side by side, the latter would appear 600,000 times brighter. One of the faintest stars known is the companion to a star called BD+4°4048. Its luminosity is only about $\frac{1}{500,000}$ that of the Sun.

The formula relating L, T, and D is

$$D = (5750/T)^2 \times \sqrt{L}$$

D = units of the Sun's diameter
T = temperature of the star in degrees absolute
L = units of luminosity in terms of the Sun's brightness
5,750 = the average temperature of the Sun's photosphere

For example, the star Sirius has a luminosity of 27 times that of the Sun and a temperature of 9,800° K. Its diameter is thus:

$$D = (5750/9800)^2 \times \sqrt{27}$$

More simply stated, the diameter of Sirius is 1.8 times that of the Sun's.

Another indirect method to measure the size of stars makes use of lunar **occultations.** In this method, the important measurement is the time it takes the limb of the Moon to occult (hide) a distant star. The bigger a star, the longer it takes the Moon to hide it. Unfortunately, this method only provides diameters for stars in the path of the Moon. In another method—**speckle interferometry**—many short ($1/100$ of a second) photographic exposures of a star are taken. Using large computers, scientists can compute the diameter of a star from these (speckled) images by removing the distortion of the Earth's atmosphere.

Stellar Mass and Density

Most stars show only small variations in mass. The majority have masses between $1/5$ and five times that of the Sun. Among the most massive stars is HD 698 with a mass 113 times that of the Sun. At the present time, there is no direct method of finding the mass of a star. There are, however, several indirect methods. One of these can only be used with pairs of stars known as **visual binaries.**

A binary is a pair of stars which, like the Earth and Moon, revolve around a common center of gravity. If the two stars of the pair are separately visible, the pair is called a visual binary. Over 50,000 visual binaries are now known. The heavier partner will describe a small ellipse around the common center of gravity, while the lighter will describe a larger ellipse according to one of Kepler's laws of planetary motion. From the sizes of these ellipses, the ratio of the two masses is determined. The sum of the masses (computed by another formula) is all that is needed to determine the mass of each individual star. For example, if the sum is 8 solar masses, and the ratio is 3 to 1, then their individual masses are 6 and 2 times that of the Sun.

Masses can also be determined for another type of binary system known as **spectroscopic binaries.** A spectroscopic binary is a pair of stars that appear as a single unit even in a large telescope. Their separation shows only in a spectroscopic study. The spectrum indicates that two stars are alternately approaching and moving away from the terrestrial observer, with their motions similar to the two masses of a rotating dumbbell. There are many thousands of known spectroscopic binaries. An example of a spectroscopic binary is Capella, a bright star in the constellation Gemini. The mass of the brighter member is 4.18, and that of its companion 3.32 times the Sun's mass, respectively. But the effective measurements of many stellar masses are questionable due to the unknown tilt of their binary orbits.

Another unique method of determining stellar masses applies to certain stars that have a very high surface gravity. Large gravitational forces are present in several kinds of stars. One class is the white-dwarf star. The high surface gravity in the white dwarf is due to the extremely high density of this type of star. White dwarfs have a fairly normal mass but are very small compared to normal stars of the same mass (hence the name "dwarf"). Typical white dwarfs have diameters of no greater than 20,000 miles, or close to the diameter of two and a half Earths.

The masses of these stars can be computed with the aid of Einstein's general theory of relativity. According to this theory, light undergoes a slight change in its wavelength when escaping from a star having a large gravitational pull. Every wavelength is slightly increased. This shift of all the wavelengths of the spectral lines toward the red is known as the "relativistic shift," and though it is extremely small, it can be measured for white dwarfs. The measured values can then be used to compute the masses of the stars that produce this redshift. Typical masses for white dwarfs are about 0.6 the mass of the Sun.

Density is determined by dividing the mass of a star by its volume. Stars are generally assumed to be spheres so their volumes are computed by using the standard formula for the volume of a sphere. Stars vary greatly in density, primarily because of their wide range of volumes. It is difficult to visualize the extreme values in stellar densities. One of the higher density stars is the Pup, the companion star of Sirius. Its mass is equal to the Sun; however, its volume is only $1/_{30,000}$ as great. Since the average density of the Sun is 1.5 times that of water, the average density of the Pup is 50,000 times that of water. A tablespoon of this star would weigh a ton! And the Pup is not the densest star known. Neutron stars have central densities of 10^{14} to 10^{15} times that of water. A cubic centimeter of a neutron star would have a mass of 10^{12} kilograms—or a million million kilograms.

At the other extreme, there are stars that have densities less than $1/_{1,000}$ of that of air. The density of these stars is even less than the density of an ordinary vacuum created in the laboratory! They are often called "red-hot vacuums." The largest known star, ϵ (epsilon) Aurigae B, also has the distinction of having the smallest density with a value of $1/_{100,000,000}$ that of water.

Stellar Motions

It is known that the fixed stars move at rather high speeds. In the course of many centuries, these movements will slowly change the shapes of the familiar constellations. The fact that these high speeds have not greatly changed the constellations is due to the great distances of most stars from the Earth. In addition, it has been a rather brief time (in astronomical terms) that the stars have been under systematic observation.

The measurement of stellar speeds requires great precision and is further complicated by the motion of the observer. Not only does the star move, but the observer, too, participates in several motions:

- Daily rotation of the Earth around its axis.
- Slight changes in direction of the Earth's axis.
- Annual revolution of the Earth around the Sun.
- The movement of the Sun and the whole Solar System in space.

These motions cause displacements of the stars called "common motions," and have nothing to do with the real movement of the stars. Common motions must be subtracted from the total displacements of stars to arrive at their actual motions.

Radial velocity, or the component of a star's speed that lies along the observer's line of sight, is determined from the spectrum of a star. The computation makes use of a basic principle in physics known as the Doppler Effect (see Chapter 7). According to this principle, the spectrum of an approaching source of light has all of its wavelengths shifted toward the blue end, while the spectrum of a light source moving away has all of its wavelengths shifted toward the red end. The same effect applies to sound waves as well. This is the reason the pitch of a train whistle increases as the train approaches a listener, then decreases as the train moves away.

In practice, photographic plates are used in this kind of work. Two spectra are photographed at the same time on one plate—one above the other. The spectrum of the star under study is obtained together with a comparison spectrum, usually that of iron. The comparison spectrum is not moving relative to the telescope, so its spectral lines are not shifted. If the star has no radial velocity, the iron lines in the stellar spectrum will exactly match the identical lines in the comparison spectrum. In the case of a star having radial velocity, the spectral lines will be displaced, and the velocity can be computed according to the Doppler formula. For example, among the largest radial velocities measured is 340 miles per second, or 547 kilometers per second, for the star CD–29°2277.

The **tangential velocity,** or cross motion, of a star is the speed of a star perpendicular to the line of sight. It is usually stated in miles per second or in kilometers per second. The tangential velocity of a star cannot be found directly but is obtained by multiplying the angular velocity of the star by its distance. It is common practice to call that angular velocity the **proper motion,** and to state its value in arcseconds per year. The greatest proper motion known is by Barnard's Star (named after its discoverer, the American astronomer Edward E. Barnard, 1857–1923). This star moves across the sky about 10.3 arcseconds each year and will move ½° (the angular diameter of the full Moon) in about 180 years. Most stars are too remote to show measurable proper motions. Out of 25,000,000 stars that have been investigated to date, fewer than ⅓ of 1 percent show evidence of proper motion.

The task of studying so large a number of stars is greatly simplified by using photographic methods together with a special kind of microscope known as a "blink comparator." (This is the same method that Clyde Tombaugh used to discover the planet Pluto.) Photographs are taken of large regions of the sky at intervals of 30 years. The two pictures are then viewed through the comparator where a special device alternately illuminates one and then the other photograph in rapid succession. Stars that have moved as much as 6 arcseconds in those 30 years seem to blink, while all the others remain motionless.

Knowing the radial and tangential velocities, it is easy to find the actual velocity of the star with the aid of the Pythagorean theorem (Figure 9.5). Among the highest known actual velocity of a star is 660 km/sec (410 miles per second), though most stars have actual velocities less than 100 km/sec.

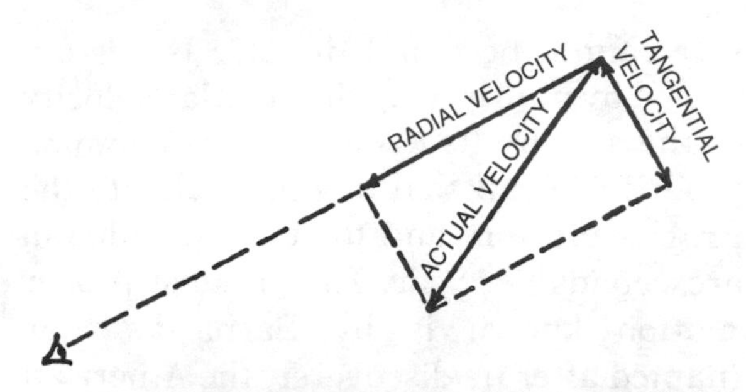

Fig. 9.5. Actual velocity. (a) The velocity of the star in line of sight, i.e., radial velocity, is determined with the aid of Doppler's formula. (b) The velocity perpendicular to the line of sight, i.e., tangential velocity, is obtained by multiplying the angular velocity of the star by its distance. (c) The actual velocity of the star in its motion through space given to scale, by the diagonal of the rectangle having radial and tangential velocity for sides.

Stellar Spectral Classes

When the spectra of many stars were analyzed, it was found that they can be naturally grouped into a number of distinct classes. The present classification is based on extensive research at Harvard College Observatory. At the end of the last century, Edward Charles Pickering (1846–1919) and his colleagues conducted a comparative spectra study of more than 300,000 stars. Historically, the classes were originally alphabetical in order. Today, ten distinct classes are recognized and given the letters O, B, A, F, G, K, M, R, N, S. It was discovered that the classes really represent a temperature sequence and were rearranged to provide a scheme of decreasing temperatures, ranging from the hot O stars to the cool M and later stars. These letters have been chosen rather arbitrarily and an easy way to remember them (as have so many astronomy students) is by the jingle: "Oh, Be A Fine Girl (Guy), Kiss Me Right Now." The last letter has several versions: "Smack," or "Sweetheart." (Three examples of stellar spectra are shown in Figure 9.6.)

Subdivisions of each of these classes are recognized and are designated by ten smaller divisions, as B0, B1, . . . , B9, and so on. K5 stands for a spectrum with characteristics halfway between K and M, for example. Our Sun is classified as a G2 star (see Table 9.1.).

The R, N, and S-type stars are a branch of the spectral sequence. These stars have chemical make-ups slightly different from the rest and are all relatively cool.

Fig. 9.6. Sirius is a class A star. Spectra emitted by these stars have: (a) wide lines due to hydrogen; (b) no line due to helium; and (c) very few and very thin lines due to metals.

Capella and our own Sun are class G stars. Spectra emitted by these stars contain many lines due to iron and other metals. Two lines due to ionized calcium are very prominent, while lines due to hydrogen are much less outstanding than in class A stars.

Betelgeuse is a class M star. Low temperature lines are strong—that is, spectral lines that can be produced by low temperature sources of light are prominent in this spectrum. High temperature lines are either very weak or entirely missing. Whole bands of lines due to titanium oxide are present in the spectra of class M stars.

TABLE 9.1 SUMMARY OF SPECTRAL SEQUENCE

Spectral Class	Temperature Range	Example
O	Greater than 25,000° K	λ (lambda) Cephei
B	11,000 to 25,000° K	Rigel
A	7,500 to 11,000° K	Vega, Sirius, Altair
F	6,000 to 7,500° K	Polaris
G	5,000 to 6,000° K	Capella, Sun
K	3,500 to 5,000° K	Arcturus
M	Less than 3,500° K	Betelgeuse

One of the incidental but important uses of this classification is to determine stellar temperatures. The variation from class to class is due, to a very large degree, to the surface temperature of the star. In fact, the surface temperature of any star is easily determined by a glance at its spectrum.

CHAPTER TEN

The Hertzsprung-Russell Diagram and the Evolution of Stars

KEY TERMS FOR THIS CHAPTER

absolute magnitude	*event horizon*	*planetary nebula*
apparent magnitude	*helium flash*	*pulsar*
black dwarf	*instability strip*	*singularity*
black hole	*Main Sequence*	*solar constant of radiation*
Chandrasekhar mass	*nebula*	*supernova*
carbon flash	*neutrino*	*white dwarf*
eclipsing binary	*neutron star*	
erg	*nova*	

On a clear night some 2,500 stars are visible to the naked eye. With a small telescope, one can see thousands more; and the largest observatory telescopes photograph billions of stars. It is estimated that there are approximately 200 billion stars in our galaxy alone—and probably more than a hundred billion galaxies in the observable universe.

At first glance, stars appear to be randomly scattered across the sky, like grains of sand on a beach. Astronomers have grouped the stars into distinct classes, each with its own set of characteristics, and have been able to determine many stars' life histories from birth to death. In this chapter, we will examine the various classes and life cycles of typical stars.

The Hertzsprung-Russell Diagram

Early in the twentieth century, the Danish astronomer Ejnar Hertzsprung (1873–1967) and the American astronomer Henry N. Russell (1877–1957), independently organized stellar classes into what is known today as the Hertzsprung-Russell, or HR diagram (Figure 10.1). The Hertzsprung-Russell diagram is a way of showing the relationship between the intrinsic brightness (or luminosity) and surface temperature of a star. The resulting diagram

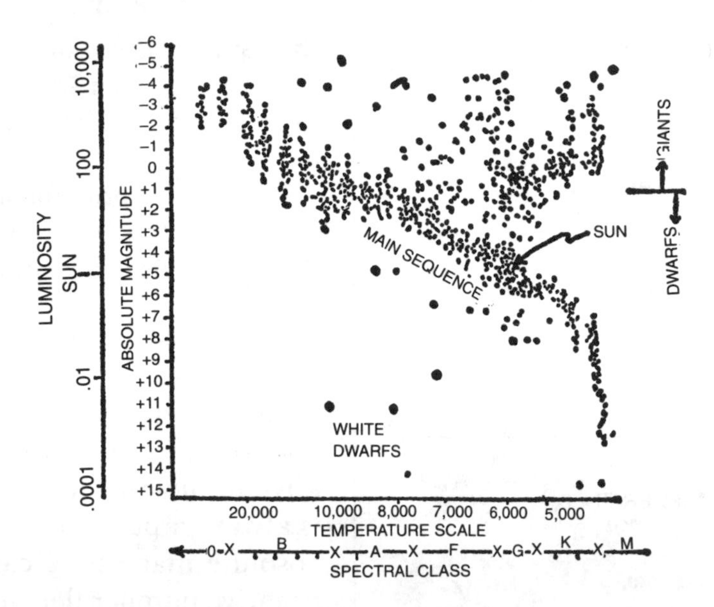

Fig. 10.1. The HR diagram for neighboring stars. On a complete diagram there are several thousand dots. Each dot represents the luminosity and the stellar spectrum of one star. The values of absolute magnitude may be used instead of the values of luminosity for the vertical scale. Similarly, the surface temperature (in ° K) of the star may be used instead of the spectral class for the horizontal scale.

The dot representing our Sun is indicated in the proper place; the Sun has an absolute magnitude of 4.8 and is in the G spectral class.

Stars brighter than +1 absolute magnitude are classified as giants; all other stars are classified as dwarfs.

shows distinct patterns and relationships that provide valuable clues about the evolution of individual stars. (Figure 10.2 shows our Sun on the HR diagram.)

The vertical axis in the HR diagram represents **absolute magnitude**, also called intrinsic brightness or luminosity. The **apparent magnitude**, or how bright a star appears, depends on a star's distance from the Earth. To understand absolute magnitude, it is necessary to eliminate the distance factor. To calculate the absolute magnitude of a star, the star is removed from its actual location to one exactly 10 parsecs (32.6 light years) away from Earth. The choice of 10 parsecs was arbitrary and is now the accepted standard.

Naturally, stars that are brought closer to the terrestrial observer will appear brighter, while stars that were pushed away to the 10 parsecs distance will now appear dimmer. The new magnitude given to a star when it is 10 parsecs away is known as its absolute magnitude. The symbol for absolute magnitude is M, while m is used to show apparent magnitude. In reality, most stars are farther away than 10 parsecs. Our Sun, however, has an apparent magnitude of −26.7 (negative numbers indicate higher luminosity than positive numbers) because it is so close. When moved to a distance of 10 parsecs, it would have an absolute magnitude of +4.8, one of the fainter naked-eye stars in the sky.

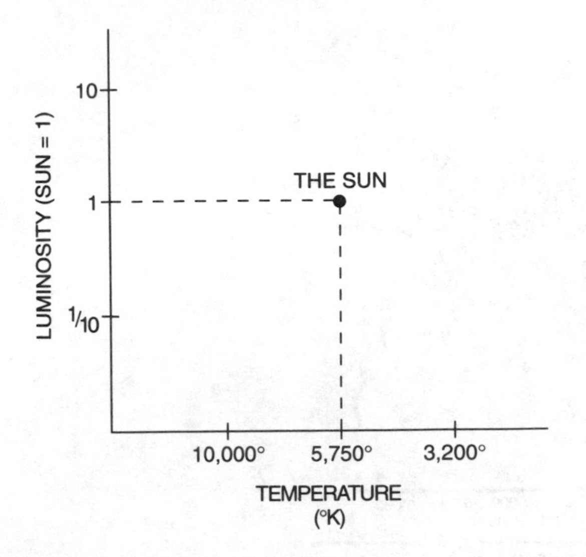

Fig. 10.2. Our Sun on the HR diagram.

The definition of luminosity is based on the total energy output of a star per unit time, or total power output. In the case of our Sun, the total power is 3.9×10^{33} **ergs**/sec. (An erg is a small unit of power in the metric system. When a fly hits a window screen, it imparts about one erg to the screen on impact. A 100-watt bulb radiates about one billion ergs per second into space.) Luminosities of stars can be conveniently stated in terms of the Sun's power output. For example, the total power of Sirius is 27 times greater than the Sun's; therefore, the luminosity of Sirius is 27 when the Sun's luminosity is expressed as 1. The symbol for luminosity is L.

Luminosity and absolute magnitude are closely related. If we know one, we can find the other with the following formula:

$$M = 4.8 - 2.5 \log (1/L)$$

where M stands for the absolute magnitude of the star and L its luminosity relative to the Sun. The luminosity of a star is always a positive number, whereas the absolute magnitude can be a positive or negative number depending on the intrinsic brightness of the star.

It should be noted that the brightness of a star can be expressed in three ways: Apparent magnitude, absolute magnitude, and luminosity. For comparison, the values of these measurements are given for five stars in Table 10.1.

This table shows that Sirius is the brightest star of the group (most negative apparent magnitude). However, if these stars were all placed at a distance of 10 parsecs from the Earth (to obtain absolute magnitude), Canopus would be the brightest as it has the most negative absolute

TABLE 10.1

Star	Apparent Magnitude	Absolute Magnitude	Luminosity
Sirius	−1.58	+1.3	30.0
Canopus	−0.86	−3.2	1,900.0
Alpha Centauri	+0.06	+4.7	1.3
Vega	+0.14	+0.5	60.0
Capella	+0.21	−0.4	150.0

magnitude. In addition, Canopus produces the most energy (or has the greatest luminosity) of the stars as well.

The horizontal axis of the HR diagram is listed by temperature, spectral class, or color. The temperature of a star, from extremely hot for blue-white stars to relatively cool for red stars, is actually the temperature of the star's photosphere, or surface. (See Chapter 9 for a complete definition of spectral classes.)

In the HR diagram, each star is indicated by a dot. The position of the dot is determined by the star's luminosity and spectral class. A careful plot of the dots for stars nearest to our Solar System shows two very important features of the diagram. First, the majority of the stars fit within a narrow band that runs from the upper left-hand corner to the lower right-hand corner of the diagram. This band is known as the **Main Sequence**, and the stars found within it, including our Sun, are known as Main Sequence stars.

Second, two well-defined types of stars fall far from the Main Sequence—the red giants and the white dwarfs. In the upper right-hand corner of the diagram are the red giants—stars that have a high luminosity but low temperatures, such as the red giant star Capella. The white dwarfs have low luminosity but very high surface temperatures and are concentrated in the lower left-hand side of the diagram. The Pup, the companion star of Sirius, is an example of a white dwarf.

The Main Sequence Stars

The fact that Main Sequence stars are found along a narrow band and not in a random pattern all over the HR diagram suggests there is a close relationship between them. The mature stars of the Main Sequence are similar in some of their characteristics, with their distribution along the sequence due to differences in mass—with the more massive stars being more luminous (Figure 10.3). This relationship is known as the mass-luminosity law. In fact, the mass of a star is the single most important factor for determining the star's entire evolution.

High luminosity stars occupy the upper portion of the HR diagram and are known as giants. The low luminosity stars are called dwarfs. The dividing point between giants and dwarfs is sometimes based on absolute magnitude. Stars that are brighter than absolute magnitude +1.0 are giant stars; while those that are dimmer than absolute magnitude +1.0 are dwarfs. The Main Sequence runs from blue giants to red dwarfs.

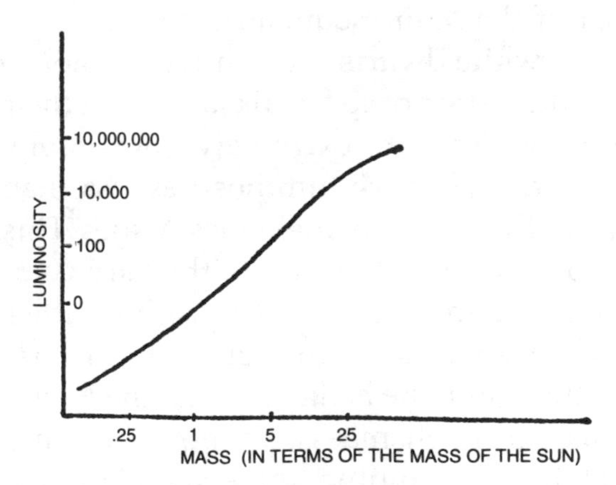

Fig. 10.3. Mass-luminosity law. High luminosity of a star coincides with great mass.

The numbers on the horizontal scale are in terms of the mass of the Sun (e.g., 5 means 5 times the mass of our Sun). The numbers on the vertical scale are in terms of luminosity (e.g., 100 means 100 times as luminous as our Sun).

Luminosity indicates the total amount of light emitted by the star into space.

Red Giants, Supergiants, and White Dwarfs

Off the Main Sequence in the HR diagram, one finds the red giants, the even larger red supergiants, and the white dwarfs. The red giants, and particularly the red supergiants, are extremely large. Some of the supergiants are big enough to accommodate much of our Solar System! The masses of some of the supergiants are only five or ten times the mass of our Sun, and because of their huge sizes, they have unusually small densities. Arcturus in the constellation Boötes is a red giant; while Antares in Scorpius, and Betelgeuse in the constellation Orion, are typical of the supergiants. The temperatures of these stars is about 2,000° K, with most of their radiation emitted in the red and infrared portion of the spectrum. These stars are more evolved than Main Sequence stars, having used up much of their fuel needed for thermonuclear reactions, and are no longer part of the Main Sequence.

The white dwarfs are similar in color to the white stars of spectral class A, but their luminosities are extremely low. Many white dwarfs have luminosities less than $1/100{,}000$ that of a normal class A star. This is not due to lack of mass (they are comparable to the mass of the Sun), but rather to their small size. In fact, white dwarfs are thought to be no larger than the Earth. Since so much mass is compressed into such a small volume, these stars have extremely high densities. A tablespoon of a white dwarf would weigh tons!

The white dwarf is one of the last stages in the life of a normal star. Our Sun is expected to become a white dwarf in another 5 billion years. Hundreds of white dwarfs are known, and there are probably many more that have not yet been detected because of their low luminosities. Scientists continue to look for white dwarfs, searching with advanced telescopes such as the Hubble Space Telescope.

There is a theoretical limit to the mass of a white dwarf, called the **Chandrasekhar mass** after the Indian astronomer Subrahmanyan Chandrasekhar (1910–) described it in 1931. He found that a white dwarf cannot be greater than 1.4 times the mass of the Sun or it would collapse into a more compact object such as a neutron star or a black hole. Chandrasekhar's work also suggested a relationship between the mass of the dwarf and its radius—the larger the mass, the smaller its radius.

Variable Stars

Main Sequence stars are relatively stable. Their masses and luminosities do not change appreciably even over millions of years. This is also true of the white dwarfs and some of the red giants and supergiants. That is why many of these stars are clumped together in these regions of the HR diagram.

There are, however, a large number of stars whose intrinsic brightness and spectral characteristics vary with time. Their changes in brightness are due to many different conditions, including two stars eclipsing each other in an **eclipsing binary** system, or from actual explosions, with the star ejecting part of its mass into space, and creating a **nova** (more on this later in the chapter). Some of the most interesting types of pulsating stars are the Cepheids, RR Lyrae variables, and long-period variables (LPVs).

Cepheid Variables

A Cepheid is a variable star whose brightness varies in a regular fashion. At the beginning of a period, its brightness increases very rapidly—for as long as several hours. This brightening is followed by a gradual dimming that may continue for several days. The cycle is then repeated and subsequent cycles are quite regular. The change in brightness from minimum to maximum is usually not very great—usually one magnitude is typical (Figure 10.4).

The first Cepheid discovered was the delta star in the constellation Cepheus (located in the northern sky just west of the North Star)—hence, the name of this class of variable star. Since the discovery of the first Cepheid in 1784, by the English astronomer John Goodricke (1764–1786), hundreds of similar stars have been discovered—with more than 600 in our galaxy alone. They are yellow supergiants having spectral classes of F and G.

Cepheids and other variable stars are named according to a fixed but confusing set of rules. The first variable discovered in any constellation is given a prefix R (for example, R Leonis); the second receives the prefix S, and so on through Z (for example, Z Andromedae). The next variable is labelled RR, and the next RS, RT, and through to RZ. Next, the letters SS, ST, and so on (for example, SS Cygni). Finally, AA follows ZZ in that scheme until QZ is reached. As no combinations with the letter J are made, this labelling method allows for 334 variables in any one constellation. Once these letter combinations are exhausted, additional variable stars are labelled by a simpler numerical method—for example, V335, V336, and V337-Cygni.

The periods of Cepheids (or the time interval between two identical points on the light curve) range from 1 hour and 28 minutes for CY-Aquarii, to 45 days and 4 hours for SV-Vulpeculae. In reality, there are two kinds of Cepheids: Cepheids I, or classical Cepheids, brighten rapidly, then dim gradually over periods ranging from 1.5 to 100 days. Cepheids II are somewhat dimmer than the first kind and are found in globular clusters in the galactic halo and in the center of the galaxy. These latter Cepheids are population II stars. (Figure 10.5 shows a light curve for the Cepheid II, W Virginis.)

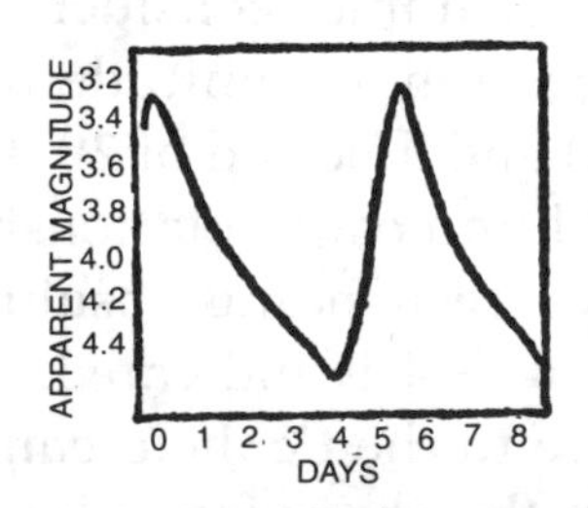

Fig. 10.4. Cepheids. The variation in the apparent magnitude of δ (delta) Cephei is shown. The maximum brightness of this particular Cepheid is 3.3; at its dimmest it has an apparent magnitude of 4.5. The complete cycle repeats every 5 days 8 hours.

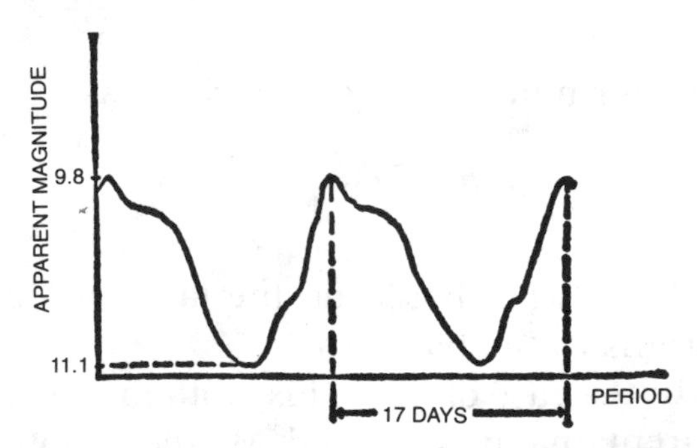

Fig. 10.5. Light curve typical of a Cepheid II star. Neither the brightening part of the curve nor the dimming part is smooth. There are characteristic leveling-off intervals in both parts of the curve. The numbers given are for the star W Virginis.

Cepheids as Astronomical Yardsticks

In 1908, at the Harvard College Observatory, Henrietta Leavitt (1868–1921) made a remarkable discovery about the relationship between the period of a Cepheid and its absolute magnitude while studying Cepheid stars in the Small Magellanic Cloud (a neighboring galaxy). Leavitt discovered that all the Cepheid stars having the same period also have the same absolute magnitude. In other words, the stars having long periods have high values for absolute magnitude; and, conversely, short-period Cepheids have smaller absolute magnitudes. This meant that all one had to do to find the absolute magnitude of a Cepheid (usually a difficult task) was to measure the period of its light variation (an easier task). This is called the period-luminosity relationship for Cepheids (see Figure 10.6).

A basic formula in astronomy links apparent magnitude (m), absolute magnitude (M), and distance (D) in parsecs. If any two of these values are known, the third value can be determined. The formula is

$$M = m + 5 - 5 \log (D)$$

The formula can also be written as

$$\log (D) = (m - M + 5)/5$$

(where log is based on the table of logarithms to the base 10).

In the case of Cepheids, both m (the apparent magnitude) and M (the absolute magnitude) are well known. The first value is obtained by averaging the light variations of the star; the second value is de-

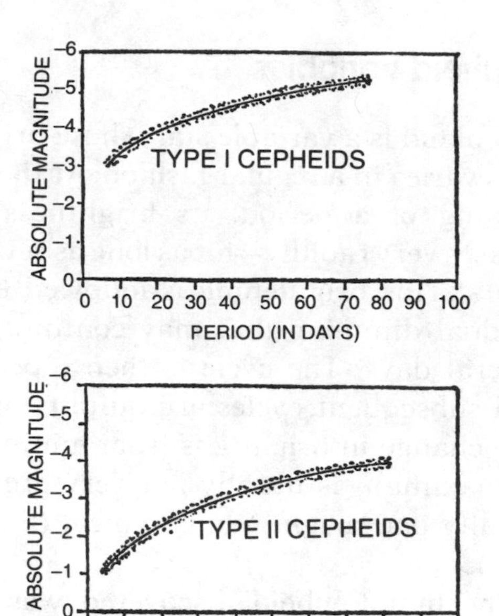

Fig. 10.6. Two curves of absolute magnitude versus period. The curve relating the absolute magnitude of a Cepheid I with its period is similar to the curve for a Cepheid II; the difference between the two curves is in brightness. For the same period a Cepheid I star is 4 times as bright as a Cepheid II. This can also be stated by saying that Cepheid I stars have values of absolute magnitudes smaller by 1.5 than Cepheid II stars having the same period of variation.

termined by using the observed period together with the proper period-luminosity relationship (there is a different one for each type of Cepheid). Astronomers then can determine the distance to the Cepheid or to any group of stars (such as another galaxy) associated with a particular Cepheid. For example, consider a Cepheid with an apparent magnitude of m = 21 (very faint) and a period of 20 days. From the period-luminosity relationship for Cepheids, a corresponding absolute magnitude of M = −4 is indicated. Therefore, the distance to the Cepheid can be calculated using the above formula:

$$\log (D) = [21 - (-4) + 5]/5 = 30/5 = 6;$$
$$\text{or } D = 1{,}000{,}000 \text{ parsecs!}$$

Hence, the Cepheids allow distances to be calculated well out beyond our own galaxy and can be used to establish the extragalactic distance scale.

RR Lyrae Variables

The RR Lyrae stars are very short-period variables, with periods averaging less than a day. The first star of this type to be discovered was a 7th magnitude star in the constellation Lyra—hence, the name of these variable stars. RR Lyrae objects were also known as "cluster-type variables" because they were first discovered in globular star clusters. This name is now obsolete because RR Lyrae variables are found in all parts of the sky—not just in globular clusters.

All RR Lyrae variables have an absolute magnitude (M) close to +0.6, independent of their pulsation period. Like the Cepheids, it is possible to use this value and the observed average apparent magnitude (m) to calculate the variable's distance in parsecs (D). Because the RR Lyrae variables are intrinsically fainter than the Cepheids, they cannot be seen at great distances and are therefore difficult to detect outside of our own galaxy.

Figure 10.7 shows the approximate locations of the Cepheids and the RR Lyrae stars on the HR diagram. These unstable and pulsating stars define the **instability strip** in the HR diagram.

Long Period Variables

Long Period Variables (LPV) are usually giant red stars with periods of longer than

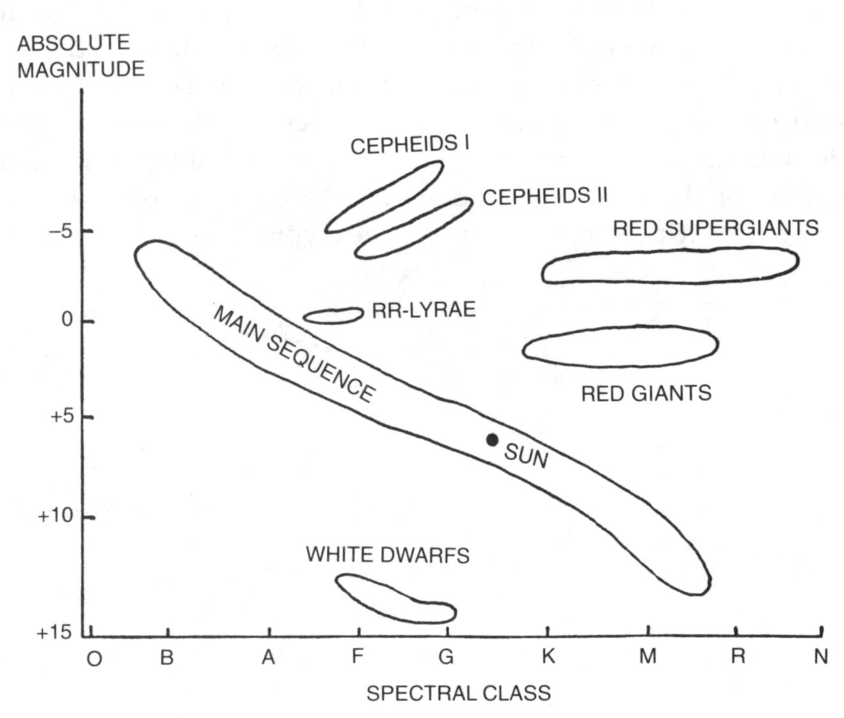

Fig. 10.7. The approximate location of the Cepheid I, Cepheid II, and RR Lyrae stars as indicated on the HR diagram.

100 days. Their light variation is not quite as regular as other variable stars. The most famous of the long-period variables is o (omicron) Ceti—discovered in 1596 and subsequently called "Mira" (the "wonderful"). At maximum brightness, Mira has been observed to reach an apparent magnitude of +1.5, making it the most brilliant star in that part of the sky. At minimum brightness, its magnitude is about +9, totally invisible to the naked eye. Its average period is 330 days, with individual periods ranging from less than 300 days to more than 350 days. Unlike the Cepheids, most LPVs do not have regular enough periods for them to serve as distance markers.

Eclipsing Binaries and Novae

An eclipsing binary is a double star system where the orbital plane is nearly edge-on to the observer. Periodically, depending on the stars' orbital velocities, one member of the pair hides (eclipses) the other. As a result, the intensity of the light from the binary varies over time. Stars in a binary system can create total, partial, and annular eclipses (Figure 10.8). By carefully studying the light curves of eclipsing binary systems, astronomers can determine the mass, size, surface brightness, and temperature of each star and the precise angle of their orbits.

Novae are stars that periodically increase in brightness and then return to their former luminosities without any apparent change in the star. Unlike other variable stars, these flare-ups are unpredictable and violent, with the average period between 30 to 50 years. The total energy output of a nova may increase by as much as 1 million times the star's prenova state over a short period of time—and consequently, a bright star appears in the sky where none was before. This is why these stars were named nova, or Latin for "new."

Novae are believed to be part of a star's late evolution on the way to becoming a white dwarf. Many of the novae that have been so far observed appear to belong to a binary system—two stars that orbit closely around a common center of gravity. Many scientists have concluded that the typical system consists of a white

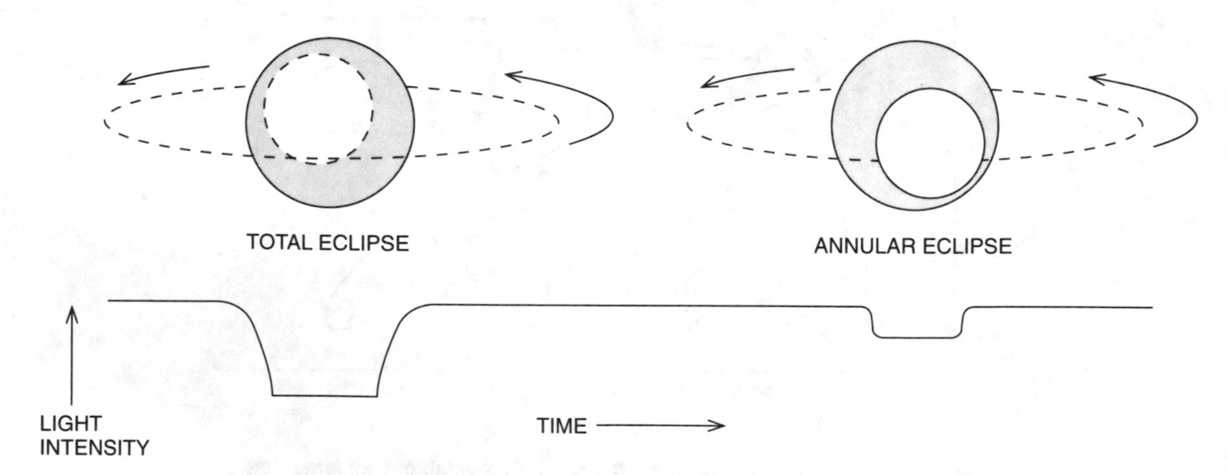

Fig. 10.8. Light curve for a completely eclipsing binary star.

dwarf orbiting a larger partner—most likely a red giant. Gravity pulls matter from the partner star onto the surface of the white dwarf. This produces instability in the white dwarf, leading to a violent outburst of X-ray radiation—a nova outburst. The newly gained material from the partner star is blown into space by this outburst and the white dwarf then settles down to its prenova state. This chain of events is frequently repeated for the majority of stars of this class.

The Energy of Stars

Stars have been emitting light and heat for several billion years. Where does all that energy come from? Many theories have attempted to answer this question. But before any theory is considered, it is useful to get an idea of the magnitude of the energies involved.

For example, the total amount of energy produced by the Sun each minute can easily be calculated. Each square centimeter of the Earth's surface that is perpendicular to the Sun's rays receives a total amount of heat and light equal to 1.94 calories per minute. This number is known as the **solar constant of radiation**. Solar energy reaches the surface of the Earth at the rate of nearly 5 million horsepower per square mile. The total amount of energy received annually by the entire surface of the Earth is truly fantastic—in excess of 5 million times the yearly production of energy obtained from coal, gas, oil, nuclear, and all other artificial sources of energy!

Knowing the distance to the Sun, it is possible to calculate the total energy radiated by the Sun per minute—472,300 billion billion horsepower. This is the energy radiated by the Sun in all directions of space, with the Earth intercepting only one two-billionths of this energy. It is believed that the energy production of the Sun has been fairly constant since it became a Main Sequence star. Therefore, it has produced this 472,300 billion billion horsepower for the last several billion years!

Before the nineteenth century, scientists were unable to formulate a satisfactory theory to explain the Sun's tremendous energy output. Ordinary combustion such as the burning of coal could not provide this amount of energy. And if the Sun were made of the best coal, it would have turned to ashes long ago. Until the beginning of the twentieth century, there was only one plausible explanation, formulated by the German physicist Hermann von Helmholtz (1821–1894) and Lord (William Thomson) Kelvin (1824–1907), his British colleague. According to their theory—often referred to as the gravitational theory because it included the idea that the Sun shrinks by its own gravity—gravitational energy was translated into heat and light energy. But it was soon shown that if this were the case, the shrinking Sun could have only lasted about 100 million years. When it was proven that the ages of certain terrestrial rocks were over 4 billion years, the foundation of the gravitational theory crumbled.

Stellar Nuclear Energy

In the early twentieth century, scientists such as Sir Arthur Eddington, Lord Ernest Rutherford, George Gamow, Hans Bethe, Charles Critchfield, and Karl-Friedrich von Weizsäcker studied the question of stellar energy production. It is now well understood that the source of the Sun's (and the other stars') energy involves nuclear reactions. Only nuclear reactions (in particular, fusion or the fusing of hydrogen into helium) are efficient enough to produce the vast amounts of energy radiated by stars for billions of years.

The conversion of mass to energy is governed by Albert Einstein's famous equation: $E = mc^2$. In this simple expression, m, measured in grams, is the amount of mass annihilated; c, measured in centimeters per second, is the velocity of light; and E, measured in ergs, is the energy obtained. For Main Sequence stars, the most abundant element, hydrogen, supplies the vast bulk of matter that is used in the nuclear reactions of the star's core. In the Sun's core, some 4 million tons of hydrogen is being converted into energy every second!

The Evolution of a Star

The life of a star can conveniently be divided into various stages:

- Birth (the local concentration of gases and dust)
- Infancy (the contracting stage)
- Maturity (moving onto the Main Sequence)
- Late evolution (the red-giant phase)
- Instability (variables, novae, planetary nebulae, and supernovae)
- Last stages (white dwarfs, neutron stars, and black holes)

The life history of a star depends greatly on its mass. The star's mass determines what class of star it will become, its life expectancy, and the kind of end it will meet. Life expectancy of stars varies from several million years for the most massive stars to many billions of years for the least massive stars. At any given moment, there are billions of stars in every stage of stellar evolution. Some have just been born, others are in the prime of life, and still others are in their declining years.

As with all theories, an assumption must be made about the origin of stars. Our starting point is a large cloud of cold gas and dust in interstellar space called a **nebula**. The average density of matter in a nebula is several thousand atoms per cubic centimeter (or 5×10^{-21} grams/c^3). The temperature is just a few degrees above absolute zero, 3° to 10°K (−270° C, or −456° F).

When the first generation of stars were formed in the very early universe, such nebulae were composed of hydrogen and lesser amounts of helium. This material was the primordial matter that had formed with the universe. The other 90 or so natural elements were produced a billion years later in the cores of very massive stars under conditions of extreme heat and pressure. These elements found their way into interstellar space largely through violent and explosive novae and **supernovae**. (Supernovae will be discussed in more detail later in this chapter.) Later generations of stars like our Sun contain small percentages of these elements in addition to hydrogen and helium.

Stellar Birth

The movements of local gravitational fields in a dust and gas nebula often create dense concentrations of matter. These concentrations eventually grow dense enough to become small, dark globules, or protostars, often seen within nebulae. Protostars remain hidden by the surrounding material and are invisible for hundreds and thousands of years—until they shine by their own light after thermonuclear ignition. The most famous stellar nursery is the Orion Nebula in the constellation of Orion, the Hunter. This cloud, rich in star-forming matter, is easily seen with binoculars or a small telescope in the northern hemisphere's winter sky.

The first generation of protostars formed some 10 to 12 billion years ago. This process continues with new generations of stars constantly being born. At birth, the star is much too cold to emit visible light, and its energy, largely in the form of radio and infrared waves, is derived from gravitational contraction.

Stellar Infancy

The large mass of nebular material, under the influence of its own gravity, shrinks and converts its potential energy into heat. The principal emissions change from radio to infrared waves. The object is now referred to as a protostar.

This shrinking and heating process takes place quickly astronomically speaking—within 30 million years—and involves three basic stages:

1. The large mass, originally with a radius of trillions of miles, decreases to only several hundred million miles.
2. The pressure at the center increases from almost zero to several thousand million atmospheres.
3. The temperature at the core rises from several degrees absolute to about 20 million degrees absolute—high enough to initiate the thermonuclear transformation of hydrogen into helium. Once this happens, the star begins its mature stage and can be plotted on the Main Sequence position of the HR diagram (Figure 10.9).

The time required for the transition from birth to maturity depends on the initial mass of the star. Massive stars develop rapidly and may reach maturity after several hundred thousand years; while less massive stars may take much longer than 30 million years to enter the Main Sequence. Massive stars naturally enter higher on the Main Sequence than less massive stars because the more mass, the more luminosity. During the millions (for massive stars) to billions (for less massive stars) of years spent on the Main Sequence, a star will move up around 1 magnitude on the HR diagram.

Stellar Maturity (The Main Sequence)

As noted earlier, nuclear energy in stars is produced from the conversion of mass into

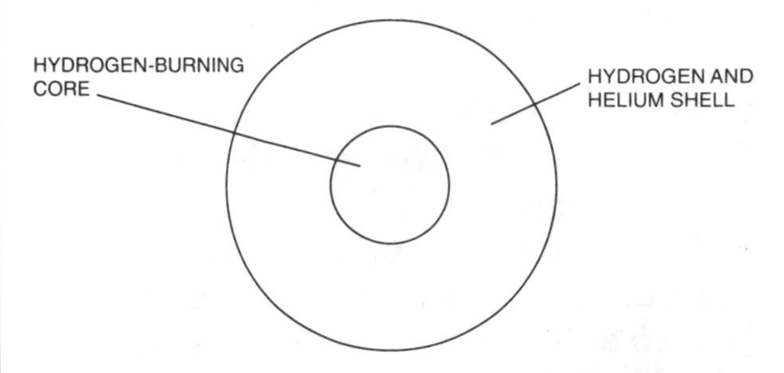

Fig. 10.9. The structure of a typical star entering the Main Sequence. The star begins to fuse hydrogen into helium.

energy according to Einstein's formula. The mass in this formula is actually the difference between the sum of the masses of the lighter atoms that go into the reaction; and the masses of the heavier atoms that are a result of the reaction.

When 4 atoms of hydrogen fuse together at high temperatures near the center of a star to form 1 helium atom, the loss of mass is 5×10^{-25} grams. From Einstein's formula this becomes

$$E = (5 \times 10^{-25})\,(3 \times 10^{10})^2 = 4 \times 10^{-5}.$$

Thus, on the Sun or any other star, 4×10^{-5} ergs of energy are produced each time 4 atoms of hydrogen fuse to form 1 atom of helium.

Our Sun emits around 4×10^{33} ergs every second. In order to produce this tremendous amount of energy every second, the Sun's core fuses 1.54 billion pounds of hydrogen and produces just slightly less helium in turn. While these numbers are enormous, they are only a fraction of the hydrogen that the Sun has available. Therefore, our Sun will most likely shine for another five billion years.

There are two distinct processes for the fusion of hydrogen into helium. One is known as the proton-proton chain and the other as the carbon cycle. The steps of the proton-proton chain are the following:

1. Two hydrogen atoms combine to form an isotope of hydrogen that is called deuterium.
2. Deuterium picks up a hydrogen atom, forming a light isotope of helium.
3. Two atoms of light helium form the final helium atom.

The carbon cycle is more complicated, with carbon, nitrogen, and oxygen alternately appearing and disappearing in the reaction. In this process, the net effect of six distinct nuclear reactions is that 4 hydrogen atoms fuse to form 1 helium atom—similar to the proton-proton cycle—with the same amount of energy released. In stars more massive than our Sun, with minimum temperatures of 10^7 °K, the carbon cycle is the predominant fusion reaction; whereas in stars like our Sun, with relatively cool cores, the proton-proton reaction predominates.

Stellar Old Age (The Red Giants)

When the core of a star runs out of available hydrogen fuel and is left with primarily helium, the star momentarily falls back on its earlier source of energy—gravitational collapse. The core begins to contract and gets hotter. As a result, the star evolves into its next phase—a red giant.

First, the temperature outside of the core rises high enough for the hydrogen to fuse into helium in a process known as hydrogen shell burning. Next, the outer layers of the star expand so that the star swells by a factor of 50. This outward expansion can no longer effectively absorb the enormous outpouring of energy that comes from the blazing hydrogen shell. Only a small part of this energy can be taken up in the expansion of the star's outer envelope; most of the energy reaches the surface and escapes as radiation. As a result, the brightness of the star increases as the surface temperature decreases. Depending on the mass of the star, it becomes a red giant or supergiant.

Simultaneous with the expansion of the star, the helium core contracts and gets hotter. At around 100 million degrees Kelvin, a new reaction begins—helium burning—and the star briefly stabilizes. Now the star's helium core furiously converts helium into carbon, oxygen, and tremendous amounts of nuclear energy in the form of gamma rays. At this point, it seems that the release of all this energy would increase the star's luminosity—just as it did when the star was burning hydrogen on the Main Sequence.

Just the opposite is true in the case of a helium burning core. Such a core has the properties of a solid steel ball because it has been greatly compressed. Because the core cannot expand, the temperature continuously rises; and, as the temperature rises, the rate of the nuclear reactions increases, raising the temperature, and the cycle continues. Thus, the core has changed into a time bomb. It takes only a few hours from the onset of helium fusion for the core to reach the explosion point. Because this time period is so short compared to other astronomical events, it is called the **helium flash**.

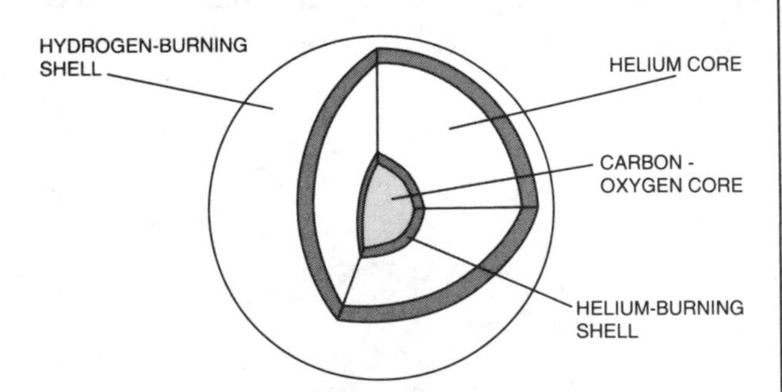

Fig. 10.10. The structure of the core of a star where a carbon core is surrounded by a shell of burning helium. The helium core is surrounded by a shell of burning hydrogen.

There are numerous events after the helium flash:

- As soon as the core explodes, its density and temperature decrease. The temperature and the rate of burning of the hydrogen shell also decrease. As a result, the star's luminosity begins to fall off and the star's distended envelope, lacking sufficient resources, begins to contract under the attraction of gravity. This descent back towards the Main Sequence continues for around 10,000 years.
- Because the helium in the core is being slowly and steadily compressed by the star's contraction, and there are no nuclear energy sources to counteract this inward force, temperatures rise sufficiently in the core to initiate a small amount of helium burning.
- By the end of 10,000 years, the core's temperature has risen to 200 million degrees Kelvin. Helium burning becomes the major source of energy for the star, with enough energy released to halt the further contraction of the star. However, now the core cannot explode because the helium core is far less dense than prior to the helium flash.
- Carbon and oxygen build up in the center of the helium core, just as helium accumulated at the early stage of the star's life. As these elements increase in number, they form an inner core within the helium core and stop the helium-burning reactions in the center. Helium burning now occurs in a shell around the carbon-oxygen core (Figure 10.10). (These events are analogous to when hydrogen was burning around an inert helium core.)

• The carbon-oxygen core contracts, and its temperature rises, which causes the helium shell to burn at a faster rate. The higher temperature of this shell is absorbed by the expanding, cool outer envelope. The temperature once again drops and the star's brightness increases.

• The star swells once more on its second climb on the HR diagram towards the red giant region. This transition takes only a few million years in contrast to the several hundred million years the star needed to reach the first red-giant stage. As the star's core becomes predominantly carbon-oxygen, it is nearing the end of its life.

Planetary Nebulae

As a star of less than four solar masses expands and cools, the outer envelope moves outward faster and faster, and eventually leaves the star. At first, the envelope is still dense enough to hide the core, and the star appears like a giant red object. Then the envelope literally blows off into space becoming a thin, nearly transparent shell of atoms that rapidly expands outward. The core of the star is now exposed and visible. The star now appears as a small, white-hot object (the core) surrounded by a softly glowing, diffuse shell of gas (the blown-off envelope). After the cool, outer envelope of gases is blown off, the star moves left on the HR diagram to the region of higher surface temperatures. This is because astronomers are now measuring the hot core (around 50,000°K) as opposed to the cool envelope (around 3,500°K).

These objects are called **planetary nebulae**. There is absolutely no connection between planetary nebulae and planets or solar systems. The name "planetary" was first used by early astronomers who, while looking through their small telescopes, thought that these objects resembled the disks of planets. The most prominent example is the Ring Nebula in the constellation Lyra—an object that resembles a smoke ring or donut.

White Dwarfs

At this point, the star begins its evolution from the core of a planetary nebula to a **white dwarf**. The star is now a carbon-oxygen core surrounded by a helium-burning shell (Figure 10.11). The core continues to contract because the temperatures in the core are not high enough to initiate fusion—so no source of energy counteracts the effect of gravity. If it were not for the electrons so densely compressed in the core, temperatures might reach the necessary 600 million degrees Kelvin to initiate carbon fusion. The electrons become incompressible when the star reaches a radius of about 5,000 miles and a density of about 10 tons per cubic inch. Therefore, the star can only radiate the energy stored in the thin helium-burning shell.

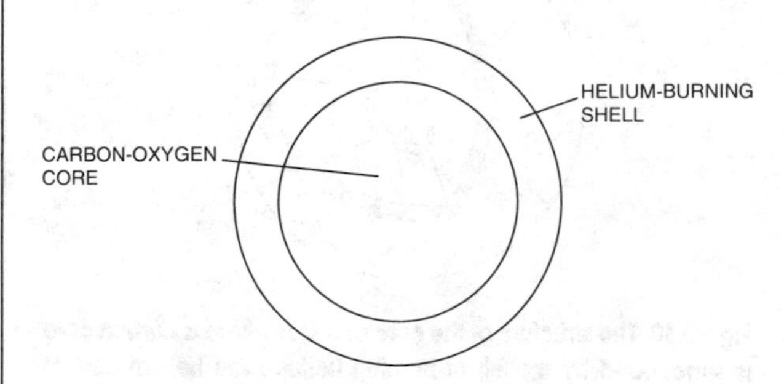

Fig. 10.11. The structure of a planetary nebula's exposed core evolving into a white dwarf.

The star is now very dim and small. If it were originally like our Sun, it would be 100 times fainter and have a diameter of only 20,000 miles—roughly twice the size of the Earth. But a tremendous amount of mass is concentrated in this small volume. A teaspoon of this star's material would weigh 10 tons! Astronomers have studied a white dwarf in the constellation Eridanus (only 16 light-years distant) and confirmed the above theoretical predictions.

A white dwarf initially has a surface temperature of around 30,000°K. As it slowly radiates this heat into space, it decreases in temperature and luminosity, and follows a path to the graveyard of stars in the lower right HR diagram. The star gradually turns from white to yellow and to red, until it finally fades out—a lump of cold, dark ash. It is now a **black dwarf**.

Supernovae

Stars greater than four solar masses end their lives in a different and far more dramatic event called supernovae. Because there is so much material in larger stars, carbon fusion begins once the temperature reaches 600 million degrees. This sets in motion a chain of events that eventually leads to the death of the star in a massive and violent explosion as powerful as the energy output of an entire galaxy.

A massive star begins its life, like most stars, by burning hydrogen to form helium. As the star evolves, it grows hotter at the center, with helium burning to form carbon and oxygen. For stars between four and ten solar masses, the buildup of heavier elements does not proceed very far. As is the case with the helium flash, the electrons in the core become incompressible and rigid like a steel ball. As soon as the temperature reaches 600 million degrees, the carbon starts to burn, but the core does not expand. This causes the temperature to rise rapidly, which makes the carbon burn faster in a process known as the **carbon flash**. A runaway effect occurs, producing temperatures and pressures that detonate the carbon core—causing the entire star to explode.

The exploding star is called a Type I supernova. All the heavier elements that were produced in each layer of the star during its lifetime are ejected into space. What is left in the star's original position is a highly compressed remnant of the core with only a fraction of the original mass.

If the mass of a star is greater than around ten solar masses, a carbon flash does not occur because the density of the core never reaches the incompressible steel ball stage. (More massive stars, have less dense cores than less massive ones. They are much more energetic, driving more material from the core toward the surface.) As the carbon-oxygen core burns, it forms heavier elements such as silicon, magnesium, sodium, chlorine, argon, and sulfur. Eventually, the core becomes so hot and dense that iron forms in the core. Iron has the special property of absorbing energy rather than emitting it. Nuclear reactions can no longer be generated—so the fire is extinguished. As a result, the pressure drops dramatically, and the core begins to contract and heat up. The rise in temperature would normally counteract the contraction, but because of the special energy-absorbing property of iron, the star continues to contract.

At this point, **neutrinos**, small subatomic particles, leave the core of the star. (In fact, recent studies of supernova 1987A in the Large Magellanic Cloud have shown that, in the last 24 hours of the star's life, energy in the form of neutrino emission may have been 10 to 100 times more powerful than the particles and radiation formed in the explosion.) This drain in additional energy from the release of neutrinos greatly accelerates the collapse of the star.

The effect of this collapse is tremendous. All the matter of the collapsing star accumulates in the center, producing extremely high pressure and temperatures around a trillion degrees. The core can no longer contract because there is no space, as all the adjacent atomic nuclei are touching each other. The star is like a giant spring. All the compressed material in the center rebounds so violently that the star explodes as a Type II supernova.

Astronomers have calculated that a supernova happens once every century in our galaxy. Yet, only three have been recorded in the last thousand years. (Others may be hidden from our view by vast clouds of interstellar dust.) Records from the imperial Chinese court noted a brilliant new star that suddenly appeared in 1054 A.D. It remained visible in daylight for 23 days and finally faded after 650 nights. In Arizona, pre-Columbian paintings of star-like objects on rocks may show the same event. Astronomers later discovered the remnant of this supernova in the constellation Taurus and named it the Crab Nebula. In 1572, Tycho Brahe recorded a supernova; and Johannes Kepler noted the last supernova in our galaxy in 1604. The 1987 explosion of a supernova in the Large Magellanic Cloud was close—around 170,000 light-years away—but still outside our galaxy.

Neutron Stars and Pulsars

Earlier in the twentieth century, theoretical astronomers Fritz Zwicky (1898–1974), Walter Baade (1893–1960), and physicists J. Robert Oppenheimer (1904–1967) and George Volkoff (a student of Oppenheimer) predicted what happened to the remnant star after a supernova. When the star explodes, the pressure in the core is so great that the electrons combine with protons to form a ball of electrically neutral particles called neutrons. According to this theory, the ball has a radius of only 10 miles but contains a large part of the star's original mass in its tiny volume. The scientists called this hypothetical object a **neutron star**.

In the 1940s, a systematic search was conducted to find a neutron star. A faint star in the heart of the Crab Nebula was tentatively identified as the supernova remnant of 1054 A.D. But there was no proof that it was a neutron star and interest in the theoretical star faded. Then by accident in 1967, a young graduate student, Jocelyn Bell (1943–), at Cambridge University, came across some unexpected fluctuations in a signal from the constellation Vulpecula. These signals showed very regular pulses, each lasting 0.016 seconds. Having eliminated all possibile terrestrial sources, Bell and her mentor, Anthony Hewish (1924–), half-seriously entertained the idea that an intelligent, alien civilization was trying to make contact with the Earth.

In order to make certain that the pulses were not just a freak occurrence, Bell scanned several other sections of the sky. She discovered three other sources with similarly regular pulses. It seemed highly unlikely that four alien civilizations were simultaneously trying to contact the Earth; but the idea persisted and the signals were listed as LGM (for "little green men") 1 through 4. Bell and Hewish thought that some form of vibration in a superdense neutron star might be the source of these radio pulses, but they could not construct a satisfactory physical model.

The pulsating radio sources were called **pulsars**. When a pulsar was discovered by the National Radio Astronomy Observatory at the heart of the Crab Nebula (where theory said a neutron star should be), the pieces of this puzzle came together. Thomas Gold (1920–) at Cornell University explained that if a pulsar was a hypothetical neutron star, it should be spinning very rapidly. He stated that as a normal star collapses, its spin rate must increase to conserve angular momentum. The analogy has often been made to ice skaters, who bring their outstretched arms down and closer to their bodies to increase their spin rate.

When the core of a supernova collapses to a diameter of only a few miles, it also drags in the star's original magnetic field where it is concentrated one billionfold at the surface of the neutron star. Plasma at the magnetic poles is whipped around with the spinning star, producing very strong radio emissions. If the Earth is in its direct path, observers will pick up this rapidly rotating radio beacon like the pulsating light from a lighthouse.

More than 450 pulsars have been discovered so far, and theory predicts that there must be thousands more in our galaxy alone. In addition, some pulsars, radiate energy in very energetic X-ray and gamma ray wavelengths. The Crab Nebula pulsar rotates 30 times per second and is one of the few that can be detected optically. Astronomers have recently found pulsars that rotate thousands of times per second. These millisecond pulsars are thought to be newly formed. For example, the first millisecond pulsar, discovered in 1982, rotates 642 times each second! Pulsars decrease their rate of rotation very slowly as they age, giving astronomers a method to measure their life spans.

Black Holes

The neutron star is no longer believed to be the final state of certain massive stars. If the mass of a collapsing star is greater than between 10 to 30 solar masses, the core of the star may be squeezed beyond the neutron star's 10 mile limit until it is no more than 2 miles in radius. At this point, Einstein's theory of relativity is used to predict a truly exotic phenomenon.

At such densities, the surface gravity of the collapsed core (or even the whole star if it was massive enough) is so great that it prevents everything, including light, from leaving the surface of the star. The gravitational force on this super-dense mass is billions of times stronger than any gravitational force found within our Solar System. Light is trapped by gravity within the star; no radiation can escape. The star blinks out and is now a **black hole**.

Black holes are virtually isolated from the rest of the universe. Nothing can emerge from a black hole. Therefore, whatever enters it remains there forever. It is impossible to know what is happening inside a black hole because the light and material particles emitted from events (or objects) inside the boundary, called the **event horizon**, cannot escape.

Once a black hole forms, its powerful gravity feeds its mass. At the center is a continuingly shrinking remnant that, according to theoretical physics, becomes smaller than anything science has ever measured. At this point the center is called a **singularity** where there is always a mass of ten thousand trillion trillion tons. As more matter enters the black hole, its gravitational pull increases, and its boundary expands. The event horizon defines the size of the black hole. Furthermore, the shrinking singularity (still with that tremendous mass) does not mean that the size of the black hole is decreasing. Astronomers estimate that black holes can range in size from that of an atom—mini-black holes—to giants with a radius of around 180 billion miles.

If black holes are invisible, how can astronomers ever detect them? Black holes can be found and studied by the effect they have on nearby stars. Some multiple stars are suspected of having a black hole as one member. For example, a powerful X-ray source is in the constellation Cygnus. The source is called Cygnus X-1 and is located near a blue supergiant whose spectrum revealed that it had an invisible companion—the source of the X-rays. If the invisible star's mass is greater than five times the Sun's, then it is a good candidate for a black hole. As gases are drawn from the surface of the blue giant to the black hole's event horizon, the gases are greatly accelerated. The rapidly moving particles collide with each other to produce a stream of X-rays like those detected by astronomers. (In 1983, another candidate for a black hole was detected in the Large Magellanic Cloud—LMC X-3.)

At first, there was skepticism about the existence of black holes because they are so different from most objects in the sky. But Cygnus X-1 and LMC X-3 seem to fit the model. Scientists are now theorizing that black holes may power every galaxy, including our own Milky Way.

The life histories of stars do not end here. The material from every star that dies is recycled into the interstellar medium. The gases and dust that form the next generation of stars, planets, and all material things (including human beings) have all been enriched with elements formed in the interiors of the previous generation of stars. When those stars died, their remains scattered to form the next generation of stars—and the cycle continues.

CHAPTER ELEVEN

The Milky Way Galaxy

KEY TERMS FOR THIS CHAPTER

dark nebulae	*globular cluster*	*Population I*
emission nebulae	*interstellar medium*	*Population II*
galactic halo	*nebula*	*reflection nebulae*
galactic year	*open (galactic) cluster*	*zone of avoidance*

The Shape of the Milky Way

The Sun, around 200 billion other stars, and a great amount of gas and dust are bound together by gravity in a very large structure called the Milky Way galaxy—or simply, the galaxy. (Other galaxies will be discussed in the next chapter.) Our galaxy gets its name from the hazy band of light, visible in dark skies, that stretches from horizon to horizon. In 1610, Galileo used his telescope to discover that the Milky Way was made up of countless individual stars.

From later studies of the distribution of stars in our galaxy, it was determined that the galaxy is shaped like a flattened disk similar to a double convex lens. Viewed from above, it appears nearly circular, with a concentration of stars in the center forming a spherical nucleus and other stars forming spiral arms. The entire structure resembles a giant pinwheel (Figure 11.1).

The diameter of the galaxy is approximately 100,000 light-years, and its maximum thickness is estimated to be between 5,000 and 15,000 light-years. Our Sun, with its system of planets, is located about 30,000 light years from the galactic center and fairly close to the galactic equatorial plane (Figure 11.2). Thus, we have a mid-

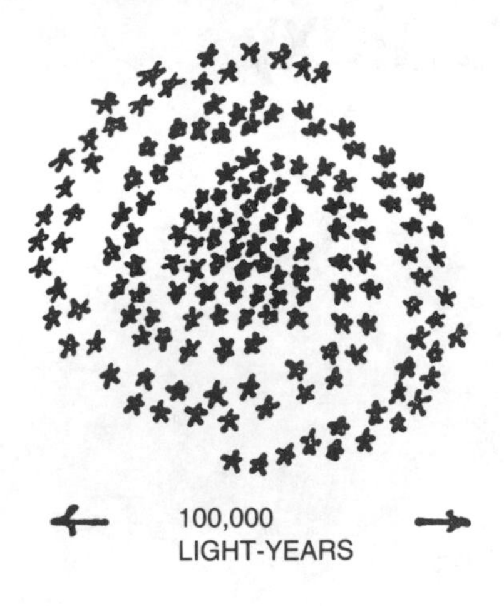

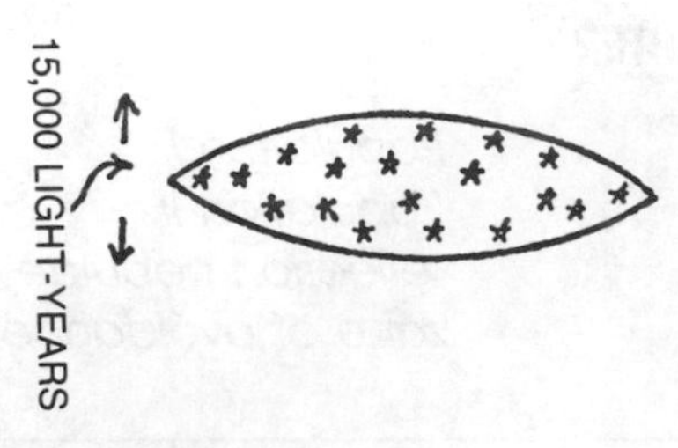

Fig. 11.1. Top view of our galaxy reveals its circular shape. The stars are not evenly distributed over that area. There is a concentration of stars in the center of the galaxy as well as along two spiral arms that start at opposite sides of the center and spiral around it.

Side view of our galaxy (bottom) shows its relative thinness.

bleacher seat view of the rest of the galaxy, with the Sun just one of many millions of stars towards the galaxy's edge.

Looking from the Earth at this vast array of stars, one gets two distinct views. In the direction towards the galactic poles, the density of the stars decreases dramatically so individual stars are seen against a dark background. Looking along the galaxy's equatorial plane, the closer stars are seen against a faint luminous band (the Milky Way)—that is, in reality, the blended light of the billions of stars in the galaxy's disk.

The shape and dimensions of our galaxy were first estimated from star counts. One of the original attempts was made in 1785 by the English astronomer (and discoverer of Uranus) William Herschel. Building a 48-inch reflector to study the heavens in greater detail, Herschel counted stars in some 683 regions of the sky. He concluded that the majority of stars are concentrated along a plane—namely, the plane of the Milky Way—and that the galaxy was disk-shaped with the Sun at its center.

This was the generally accepted view of the galaxy, until the 1920s pioneering work of American astronomer Harlow Shapley at the Mount Wilson Observatory. Shapley determined the true size and shape of our galaxy. As was noted before (Chapter 10), Henrietta Leavitt had already established Cepheid variables as a new astronomical yardstick for measuring interstellar distances. Shapley studied the Cepheid and RR Lyrae variables in **globular clusters**—swarms of stars crowded together in spherical shapes that are scattered around the galaxy (Figure 11.3). He calculated the distances to 69 of these clusters and determined that they formed

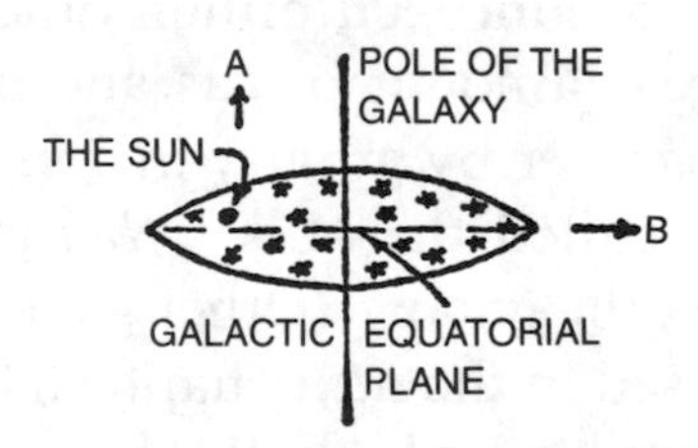

Fig. 11.2. Two views of the galaxy. Looking from the Solar System along the smaller dimension (direction A), stars are seen against a black background.

The nearby stars along direction B are seen against the background of the Milky Way. The luminosity of the latter is due to the merging points of light produced by billions of stars. The galactic equatorial plane slices the galaxy horizontally in two equal parts. The pole of the galaxy is perpendicular to the galactic equatorial plane at its center.

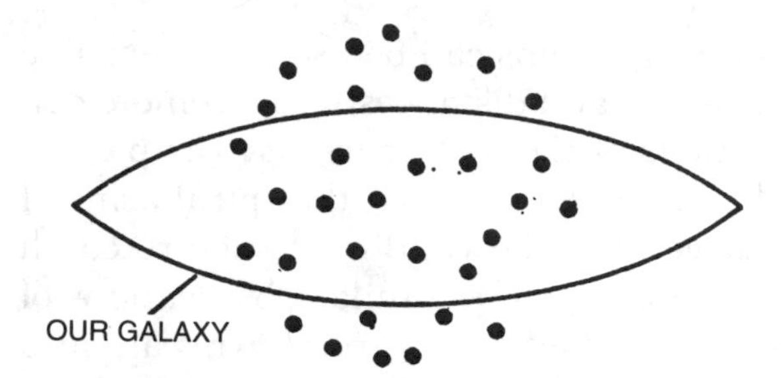

Fig. 11.3. Distribution of globular star clusters. A symbolic representation of these star clusters would show them to be distributed in the form of a sphere around our galaxy. The diameter of the sphere is equal to the diameter of the Milky Way. (One of the most beautiful globular clusters, barely visible to the naked eye, is M13 in the constellation Hercules.)

a spherical distribution centered on the constellation Sagittarius—25,000 to 30,000 light-years away. Shapley then made the brilliant deduction that if the Sun were actually at the center of the galaxy, globular clusters should appear nearly uniformly distributed across the sky. Since they were not, the center of the galaxy must lie towards Sagittarius. That places the Sun some 25,000 light-years away from the galactic center.

Rotation of the Galaxy

The shape of the galaxy implies that it is rotating. In fact, it could not exist as a flattened disk and central bulge without rotation because all the galactic matter would fill a spherical volume rather than a flattened one. (The topic of galaxy formation and the problem with spiral arms will be discussed in detail in the next chapter.) The axis of rotation is perpendicular to the equatorial plane of the galaxy. The individual stars also move within the galaxy. In this respect, it is similar to the Earth's rotation around its axis, while all kinds of movements proceed on the surface.

There is a great difference between these two rotations. The galaxy does not rotate as a solid body. The stars rotate around the center of the galaxy in much the same way as the planets move around the Sun. Stars closer to the center of the galaxy move at greater orbital speeds, obeying Kepler's laws; while stars closer to the edge of the galaxy seem to move more slowly. Relative to the Sun, these slower stars seem to be going in the opposite direction—like slower cars on the highway as seen from inside a faster vehicle.

Our Sun has an orbital velocity of 160 miles per second and takes approximately 250 million years to complete one revolution around the center of the galaxy. This figure is sometimes referred to as the **galactic year**. From the accepted age of 4.6 billion years for the Solar System, it is estimated that the Sun has made 20 revolutions around the galactic center since its birth. The last time that the Sun was in its present position, dinosaurs had just started appearing on the Earth.

Spiral Structure of the Galaxy

Although studies have shown that our galaxy is a spiral, relatively little can be seen from optical or visible light observations. This is because we are located inside the plane of the galaxy in one of the spiral arms. In addition, the effects of interstellar dust greatly limit our view of the galactic disk. It is much like being inside a forest, among many tall trees and thick underbrush—and trying to see the overall shape and design of the forest itself.

Our knowledge of our galaxy's structure has been greatly enhanced by the study of the 21-cm (centimeter) radio waves emitted by interstellar hydrogen. This idea was first formulated by a young Dutch astronomer, Hendrik van de Hulst, in 1944. The galaxy contains hydrogen throughout the entire disk, but it is denser in the spiral arms. Neutral, or atomic, hydrogen (one proton and one electron) emits radiation at a wavelength of 21 cm when the electron changes its direction of spin. In an atom of hydrogen the electron often spins in the same direction as the proton and has slightly more energy than an atom that has an electron spinning in the opposite direction to the proton. When the electron flips its spinning direction, a quantum of electromagnetic energy is given off with a characteristic wavelength of 21 cm (Figure 11.4).

This radio emission penetrates the interstellar dust (unlike visible light) and can be detected over the entire disk of the galaxy. Using sensitive radio telescopes, astronomers have mapped out the 21-cm emissions in many areas of the galaxy. Assuming a model for how the galaxy rotates,

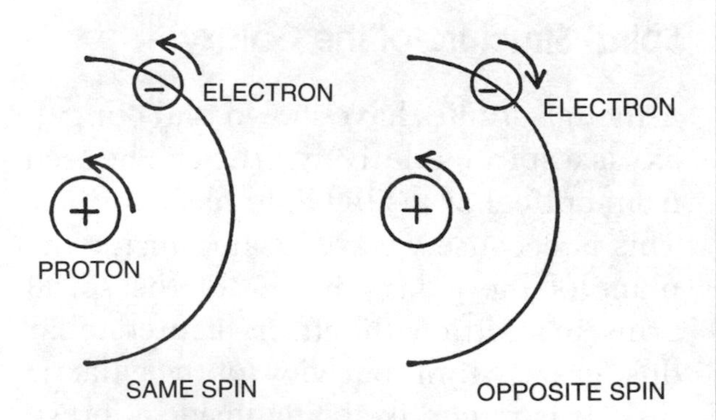

Fig. 11.4. Electron flip. On the left side of the illustration, the electron and the proton are both spinning in the same direction (counterclockwise). After the electron flips, electron and proton spin in the opposite directions.

these emissions can be used to determine densities and distances to the remote collections of the hydrogen gas. The picture that emerged revealed the spiral arms of our galaxy. The validity of such research was confirmed by similar observations of other galaxies where spiral arms are also visible—and the hydrogen emission lines matched the images on the photographs.

Mass of the Galaxy

Filling the galactic disk are between 100 and 200 billion stars, wide lanes of gas and dust, giant interstellar clouds, high-energy cosmic rays, magnetic fields, and radiation in all wavelengths from gamma rays to radio waves. Astronomers used to believe that the bulk of the galactic mass was concentrated in the disk, but recent evidence shows that most of the galactic mass resides in the **galactic halo**—a huge shell of invisible or dark matter that surrounds the galaxy. Evidence for the halo was derived from measurements that showed that the entire galaxy was rotating faster than predicted by Kepler's law. Only a tremendous mass could account for the faster rotation. The exact nature of this mass still remains unknown. At present, it is estimated that our entire galaxy contains material equal to about 1,000 to 2,000 billion solar masses.

Nucleus of the Galaxy

What do we see when we look towards the center of the galaxy in the direction of the constellation Sagittarius? In visible light, we see huge fields of stars interspersed with dark, obscuring patches of dust. The dust is so dense that it actually prevents any direct optical observations of the galaxy's nucleus. (An examination of this interstellar medium, as well as of star clusters and nebulas, follows later in this chapter.) By looking at other spiral galaxies, such as the neighboring Andromeda galaxy, we can reasonably determine what our own nucleus looks like—a nearly spherical collection of stars, surrounded by dust and gas.

Observations made in other wavelengths such as radio, infrared, X-rays, and gamma rays have allowed astronomers to pierce the curtains of dust and probe the nature of the galaxy's heart. At the center of the galaxy is one of the strongest radio sources known—Sagittarius A* (pronounced "Sagittarius A star"). Sagittarius A* is very active, much like similar objects observed in other galaxies. Studies of the nucleus in other wavelengths have revealed that perhaps 1 million solar masses of hot gas have been ejected at high speeds from the galactic center; in addition, as many as 5 million solar masses may be crowded within just a few light years of the nucleus. There is even speculation that a giant black hole is the engine that drives the galaxy.

Star Clusters

As was noted in the previous chapter, the shape and scale of our galaxy was largely determined through the study of variable stars in globular clusters. These groups of stars are born about the same time and held together by the mutual gravitational attraction of their members. The galaxy contains actually two distinct types of clusters: **open**, or **galactic clusters**, and globular clusters. These two types of clusters are very different in appearance and each represents different stages in the evolution of their member stars.

Open clusters are loosely assembled, young stars only found along the galactic plane. Some are rather inconspicuous objects containing as few as 20 stars that barely stand out against the starry background; while others are very rich clusters containing as many as 1,000 stars. Two very prominent open clusters are the naked-eye Pleiades in Taurus and the Wild Duck Cluster in Scutum. The diameters of open clusters range from about 15 to 40 light-years. It is generally believed that over time, the stars of an open cluster will separate.

Globular clusters, on the other hand, are very old objects that are found primarily in the galactic halo around the galaxy's nucleus. As many as a million stars are packed into a ball so compact that individual stars are impossible to resolve even with the most powerful telescopes. The member stars of globular clusters, having been formed from primordial hydrogen, follow a very different evolutionary track than the stars formed from the heavy element-enriched matter of the galactic plane.

The diameters of globular clusters range from about 75 light-years to as much as 400 light-years; while their distances from the Sun range from about 30,000 light years to as much as 300,000 light years. The most prominent globular cluster is M13 in the constellation Hercules—often visible as a fuzzy star to the naked eye.

Walter Baade, a German astronomer who lived in exile in the United States, made spectroscopic studies of the stars in the Andromeda Galaxy (thought to be a near twin of our galaxy). He identified two different populations of stars that he called **Population I**—young, blue stars rich in heavy elements, found primarily in the spiral arms—and **Population II**—old, red stars of primordial hydrogen, found primarily in and around Andromeda's central bulge. From his studies, he concluded that the globular clusters found in the halo of the Milky Way also formed from primordial hydrogen, and because they are so old, there is no longer any dust and gas clouds to supply a new generation of stars. Therefore, no new stars are being created there. These ideas helped astronomers determine that our galaxy is only a few hundred million years younger than the Big Bang.

The Interstellar Medium

Stars are quite enormous, but the vast spaces between them are even larger. Interstellar space appears quite empty by terrestrial standards. Yet very tenuous matter consisting of gases and minute dust particles is found throughout interstellar space and is known as the **interstellar medium**. The gas and dust, however, are not distributed uniformly. They are concentrated as dust clouds, or nebulae, that are only visible against a background of stars or detected by the telltale energy they emit. Looking up at the Milky Way stretched across the sky, it is easy to see many dark, patchy regions. These voids and gaps in the smooth background of fainter stars are the dark dust clouds of the interstellar medium.

Spectroscopic and radio emission studies have revealed that the gas is mainly atomic hydrogen (H), with smaller amounts of molecular hydrogen (H_2), carbon monoxide (CO), hydroxyl (OH), and minute additions of other elements like carbon, nitrogen, oxygen, sodium, iron, and titanium. The composition of the dust particles is less well known but may consist of a large number of carbon atoms joined together to form graphite. Some studies show that the dust particles may also contain combinations of icy crystals of water (H_2O), ammonia (NH_3), and methane (CH_4). (It is interesting to note at this point how similar the atmospheres of the outer planets are to this medium.)

The density of that interstellar matter is extremely low—about one atom of hydrogen per cubic centimeter and about 100 to 1,000 dust particles per cubic kilometer. Even in the thickest parts of the clouds, the density is still so low that it resembles a high vacuum by Earth-based laboratory standards. The dust is only about 1 percent of the total interstellar medium, but it is the main agent that causes interstellar reddening.

The Reddening of Starlight

The interstellar medium is responsible for both the dimming and reddening of distant stars. Starlight is dimmed in almost all directions. This can be observed by simply counting the number of stars at fainter and fainter magnitudes in various lines of sight. If a uniform distribution of stars is assumed, fewer stars are counted than should be there—evidence is that the stars are dimmed by dust. In fact, dust obscures our view so much that we cannot see further than about 15,000 light-years in the direction of the galactic plane.

It has also been found that the distribution of other galaxies is not uniform near the galactic plane. The obvious explanation for this **zone of avoidance** is that the interstellar medium concentrated along the Milky Way is hiding the light from distant galaxies.

In addition to dimming, interstellar dust also reddens starlight. The reddening effect is due to the fact that interstellar material scatters shorter wavelengths of light (blue) much more efficiently than it does the longer wavelengths (red). Hence the blue light is scattered and the red light goes through (much in the same fashion that the Earth's atmosphere creates red sunsets as sunlight travels through the densest part of our atmosphere). Therefore, the light we see from a star that is shining through clouds of interstellar dust is preferentially redder. This phenomenon explains why some stars show spectral characteristics of hot, blue type O or B stars, yet appear in color as type F or G stars.

Nebulae

In many parts of the galaxy there are dense concentrations of interstellar matter. This matter is called a **nebula**, Latin for "cloud," and should not be confused with planetary nebulae. (See Chapter 10.) The average density of material in a nebula is more than a 1,000 times the density of the interstellar medium. Nebulae are classified as **emission nebulae**, **reflection nebulae**, or **dark nebulae**.

An emission nebula is a cloud of material that has one or more extremely hot, luminous type O or B stars embedded in it. The ultraviolet light from the stars excites the hydrogen and oxygen that give these nebulae their characteristic greenish-yellow and red glows. An excellent example of an emission nebula is the Great Nebula in the constellation Orion, where very young, hot stars excite the gases left over from their formation. Two other colorful examples are the Lagoon and Trifid Nebulas in the constellation Sagittarius.

If the star (or stars) embedded in the nebula is cooler than a type B_1 star, the gas in the cloud will not be excited enough to glow. The dust in the cloud will reflect the light (usually bluish) from the nearby star and show the same spectral characteristics. Hence, this type of cloud is called a reflection nebula. Reflection and emission nebulae often form in the same cloud of material because their populations of stars are so diverse. The hotter stars cause the gas to glow; while the dust reflects the light of the cooler stars. A good example of reflection nebulae are the wisps of blue nebulous clouds surrounding the Pleiades in the constellation Taurus.

If the nebula has no nearby star to supply it with light, it is known as a dark nebula. These appear as dark patches, streaks, or globules. The most famous and striking example is the Horsehead Nebula in Orion. Other prominent dark nebulae are found in the Trifid, Rosette, North American, and Eagle nebulae. American astronomer, Edward Emerson Barnard (1857–1923), specialized in studying dark nebulae. Several are named after him, such as Barnard 86 in Sagittarius, and Barnard's Loop, just east of Orion's belt.

The Origin of Interstellar Matter

Where did the interstellar matter come from and where will it go? This is a fascinating question because it relates to the formation not only of our galaxy, but to the formation of the Sun and its planets as well. To some extent the material between the stars is left over from the formation of the galaxy and the first generation of stars. However, there is good evidence that a large percentage of this material is continually ejected into interstellar space by stars entering the late stages of stellar evolution (red giants, novae, supernovae, and white dwarfs).

The original composition of the interstellar medium was largely hydrogen with a small percentage of helium—the so-called primordial elements. We now measure trace amounts of heavier elements like carbon, nitrogen, and oxygen produced by nuclear reactions in the stellar cores. Hence, the first generation of stars returned material back into the interstellar medium; the second generation of stars, like our Sun, contains a more diverse mixture of elements. Since our own bodies are made of these elements, we can infer that we also contain material that was originally made in the fiery cores of the previous generation of stars.

CHAPTER TWELVE

Other Galaxies

KEY TERMS FOR THIS CHAPTER

cannibal galaxy
elliptical galaxy
gravitational lensing
irregular galaxy
Local Group
quasar
ram pressure stripping
relativistic velocity
Seyfert galaxy
spinars
spiral galaxy
voids
wave density

Our galaxy is not alone. The universe probably contains more than a hundred billion galaxies. Astronomers consider galaxies the basic building blocks of the universe—just as stars are the building blocks of a galaxy. And they study the structure and evolution of galaxies to gain a better understanding of the large-scale structure of the universe.

Before their true nature was determined, galaxies were often referred to as spiral nebulae because they resembled the fuzzy clouds of dust and gas found within our own Milky Way galaxy. American astronomer Edwin Hubble (1889–1953) discovered that galaxies were large families of stars, gas, and dust that varied in shape and size and were far beyond the boundaries of the Milky Way. (Another great discovery made by Hubble about galaxies will be discussed in the next chapter.)

Some galaxies are close to us. For example, the Large and Small Magellanic Clouds are about 170,000 light years away. On the other hand, some galaxies have been detected several billion light years away. In addition, galaxies have been shown to combine into larger and larger structures called clusters and superclusters. Current research indicates that the large-scale structure of the universe resembles a sponge—with strings of galactic superclusters surrounding large voids of empty space.

The Hubble Classification of Galaxies

From 1918 to 1938, Edwin Hubble studied more than 600 galaxies using Mt. Wilson Observatory's 60-inch and 100-inch telescopes. He found that galaxies come in a variety of shapes and could be classified as three major types: Elliptical (E), spiral (S, where a barred spiral is a variation, SB), and irregular (Irr).

Elliptical galaxies resemble partly flattened, luminous disks. Some are nearly spherical, while others show varying degrees of flattening. (These designations only apply to the galactic shapes as seen from the Earth.) Hubble designated a galaxy with little or no flattening as E0 and one with the most flattening as E7. Galaxies with intermediate shapes are designated E2 through E6 (Figure 12.1).

Elliptical galaxies show no evidence of spiral arms. The most characteristic property of this group is the lack of any star-forming regions. Careful observation has shown that elliptical galaxies contain only very old stars. There are none of the young blue giants found in other types of galaxies. One explanation is that all of the dust and gas in an elliptical galaxy have already condensed to form stars. Elliptical galaxies may have been swept relatively clean of dust and gas by their rapid passage through the intergalactic gaseous medium—material that acts like a powerful wind. (A more detailed discussion of the intergalactic medium follows in this chapter.)

Elliptical galaxies accounted for about 20 percent of the galaxies that Hubble surveyed. However, this number is actually too low, as many ellipticals are small and faint. Even the relatively large examples are dim in proportion to their masses, especially since there are no bright, young stars in these galaxies. Some examples of elliptical galaxies are M87 in the constellation Virgo (that may contain a monster

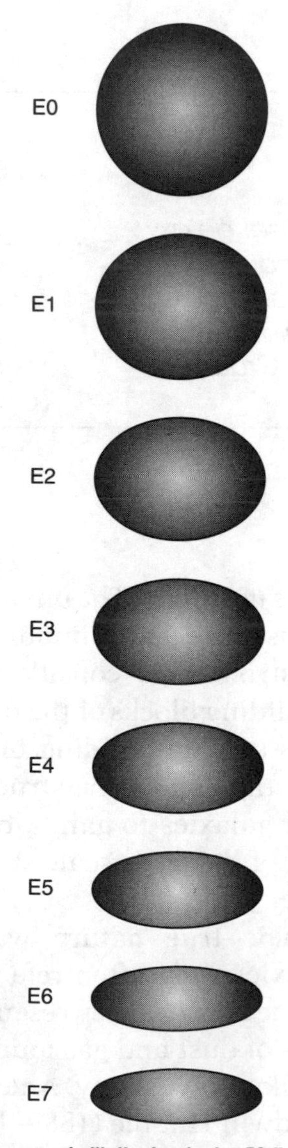

Fig. 12.1. The shapes of elliptical galaxies E0 through E7.

black hole at its center), and the two satellite galaxies of the Andromeda galaxy (M31)—M32 and NGC 205, both easily seen with binoculars or small telescope.

Hubble then classified 50 percent of the galaxies he observed as **spiral galaxies**. Like ellipticals, spiral galaxies resemble flattened luminous disks; however, the majority of the stars are concentrated in the spherical bulge, or nucleus. To a cosmic observer watching for hundreds of millions of years, the entire structure would look like a giant rotating pinwheel. (Of course, because of the galaxy's tremendous size, an observer does not see any movement within the short human lifespan.) Hubble further classified spiral galaxies into three subgroups: Sa, Sb, and Sc, according to the amount of material in the arms relative to that of the nucleus, and to the degree of tightness of the arms (Figure 12.2).

Spiral galaxies are large and massive, containing an average of 10^{11} stars in all stages of evolution. Surrounding spiral galaxies are scattered isolated stars and globular clusters (Chapter 11). An average spiral galaxy contains several hundred globular clusters spread evenly around the disk in a region known as the galactic halo. Recent discoveries about the halo surrounding the Milky Way (Chapter 11) indicate that the bulk of the material in spiral galaxies is not in the disk, but in the invisible (yet unknown) matter in the huge extended halo. (The cosmic implications of this "hidden" mass in galactic halos will be discussed in Chapter 13.)

Our galactic neighbor, M31, in the constellation Andromeda, is 2.2 million light-years away. It is an excellent example of a spiral galaxy and is believed to resemble our own Milky Way. Two other spiral galaxies are M33 in the constellation Triangulum, and M51, the Whirlpool Galaxy in the constellation Canes Venatici. Both of these galaxies are seen face on—as if an observer were looking down from one of the galactic poles.

Hubble classified about 30 percent of his sampling as barred spirals. He created three similar subgroups that follow the same general scheme for classifying regular spirals: SBa, SBb, and SBc (Figure 12.3). Barred spirals are similar to regular spirals in every respect, including their stellar populations, mass and luminosity, halos, and distribution of gas and dust.

Hubble observed a small fraction of the galaxies that did not fit into any of these categories. They showed no regular geometric form and were generally smaller, with only about a thousandth of the mass of a typical elliptical or spiral galaxy. All were rich in clouds of glowing ionized gas and were busy forming young blue stars. Hubble classified this type of galaxy as an **irregular galaxy**. One of the best known

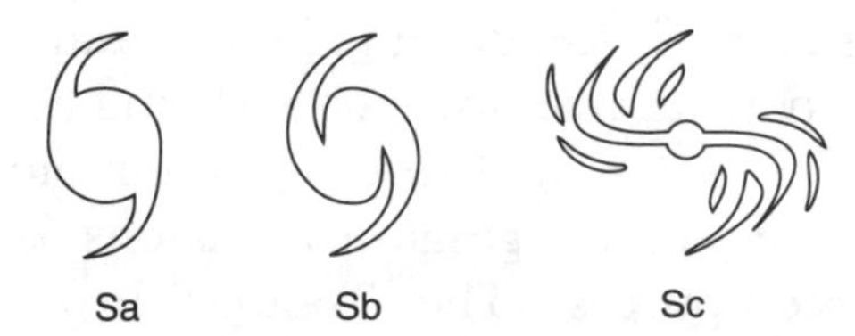

Fig. 12.2. The shapes of typical spiral galaxies.

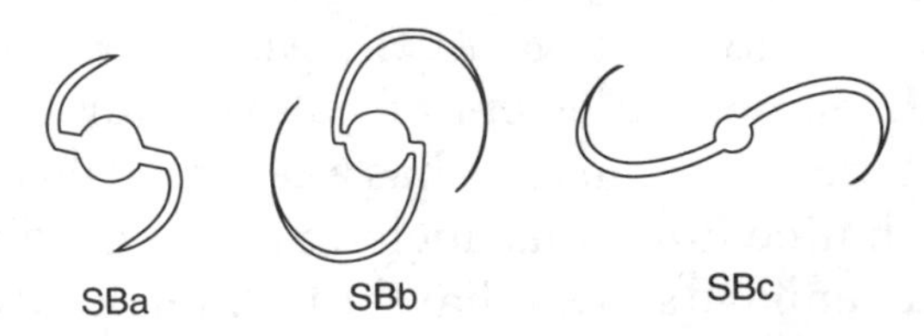

Fig. 12.3. The shapes of typical barred spiral galaxies.

and closest examples is the Small Magellanic Cloud—a dwarf galaxy that is 180,000 light-years away and a satellite of the Milky Way.

As an ironic footnote to Hubble's work, the most numerous and common type of galaxy found in the universe—the dwarf galaxy—was never formally catalogued by the great astronomer. These galaxies are very difficult to detect because they are not as obvious or luminous as the bigger galaxies—but they far outnumber them. Large numbers of dwarf galaxies have been found in neighboring clusters of galaxies. Some dwarf galaxies have regular shapes such as the Large Magellanic Cloud, which some astronomers consider a very loose, dwarf elliptical galaxy. Others, such as the Small Magellanic Cloud and the dwarf galaxy in the constellation Sextans, have no regular shape.

Cannibal Galaxies

Some of the largest galaxies in the universe may have actually consumed their neighboring galaxies. These are called **cannibal galaxies**. Galaxies designated as cD galaxies are supergiant ellipticals with several nuclei within them. These galaxies are generally found at the center of galactic clusters where the probability of galactic collisions and cannibalization are more common. One galactic cluster, Abell 407 (named for George Abell, who surveyed and mapped clusters of galaxies for the Palomar Sky Survey), has a central galaxy with nine luminous nuclei. Studies of this cluster have shown that the nine nuclei are from cannibalized elliptical galaxies.

Active Galaxies

For years, huge, slow-moving galaxies were considered to be placid and virtually stable. However, recent studies using both radio telescopes and computer models have presented a very different view. All galaxies emit small amounts of energy in the form of radio waves—but some galaxies are sources of extremely powerful radio emissions.

In 1940, Grote Reber, a pioneer in radio astronomy, detected a strong radio source toward the constellation of Cygnus, the Swan. He discovered a galaxy one billion light-years away emitting radio signals that were several million times more powerful than emissions from our own galaxy. Photographs made with Mount Palomar's Hale Telescope revealed Cygnus A—a strange-looking, double-lobed galaxy.

Debate raged between two great astronomers of the time—Walter Baade and Rudolph Minkowski (1895–1976)—about the true nature of the tremendous radio emissions. Baade insisted that the emissions were a result of galactic collisions; Minkowski believed that the energy resulted from violent activity in the galaxy's nucleus. Minkowski's interpretation eventually prevailed, but the debate has not been resolved. More recent and refined observations have recorded a complex structure in Cygnus A—evidence that could confirm some aspects of the collision theory.

The 1970s brought about a revolution in ideas about interacting galaxies. Alar and Juri Toomre, two brothers working at the Massachusetts Institute of Technology, modeled realistic galactic collisions using supercomputers. The Toomre brothers also proposed that elliptical galaxies are a pile of stars left after the collisions of spiral galaxies. Their idea was first greeted with skepticism, but there is now growing support among astronomers for their collision hypothesis. In addition, the Infrared Astronomy Satellite (IRAS), launched in the early 1980s, showed evidence that starburst activity, or increased amounts of heat from possible accelerated star formation, results from the supersonic collisions of galactic gas clouds.

Since Reber's first discovery, many more strong radio sources have been detected. Some of these galaxies, such as NGC 2535/36 and NGC 4038/9, seem to be galaxies in collision. Other galaxies, such as M82 in the constellation Ursa Major and M87 in Virgo, may derive their powerful radio emissions from titanic explosions in their nuclei. This violent activity produces long jets of matter that shoot out into space from the center of the galaxy.

Another example of a powerful radio galaxy is NGC 5128, or Centaurus A. It is a very luminous body that resembles a normal elliptical galaxy bisected by a broad, dark dust lane. Radio maps reveal two large radio sources on either side of the visible galaxy. (These twin-lobed features are present in many typical radio galaxies.) Currently, scientists believe that these radio emissions mark the boundaries of violently ejected gases from the galaxy's core.

Seyfert Galaxies

The **Seyfert galaxies** are named after Carl Seyfert, who discovered their unusual properties in 1944. This type of galaxy differs from normal galaxies in several intriguing ways:

- The total output of energy at all wavelengths greatly exceeds that of an ordinary galaxy (like the Milky Way) by a factor of 100.
- The source of this energy is a very small and bright nucleus usually less than a light year in diameter. Only when photographed with long exposures do the spiral arms of a Seyfert galaxy become visible.
- The energy output from the nucleus sometimes varies by a factor of two or more times—in a period as short as a few hours or even minutes!

NGC 1068 and NGC 4151 are two examples of Seyfert galaxies. Many Seyfert galaxies are very powerful X-ray sources, leading astronomers to believe that black holes may lie at the centers of the galaxies. Their energy output is too great to be from the ordinary stellar nuclear reactions. Yet Seyfert galaxies are not the most powerful beacons in the sky.

Quasars

In 1960, the discovery of quasi-stellar radio sources, or **quasars** (a term coined by astronomer Hong-Yee Chiu in 1964), challenged astronomy's views of the universe. For many years, these objects were found on photographs as very faint, bluish stars and were assumed to be members of our own galaxy. Around 1960, astronomers discovered that these stars were sources of strong radio emissions. They knew galaxies and nebulae give off radio signals, but how could an ordinary star generate such strong radio waves? In addition, the visible light spectra from these objects were not only unusual, they had never been seen before! Had astronomers found some new and exotic substance? Did the basic laws of physics need to be rewritten?

In 1963, Maarten Schmidt, an astronomer at the Palomar Observatory, found the answer. While studying one of these star-like objects, 3C273, he realized that the normal spectral lines representing hydrogen had shifted toward the red end of the spectrum. Astronomers had observed similar red-shifts in the spectra of stars and galaxies as the objects moved away from the Earth and approached **relativistic velocities**, or the speed of light. This shift toward the red end of the spectrum in rapidly receding light sources is known as the Doppler Effect (see Chapter 7).

Assuming that a red-shift meant that the object was moving away from the observer, Schmidt calculated that 3C273's distance was about 2 billion light years and was moving away at the incredible speed of 28,400 miles per second. From its apparent magnitude (m = +13.6) and its distance, it was calculated that 3C273 has a luminosity of 10^{46} ergs/sec, or roughly 50 times the brightness of our own galaxy. Similar values have been obtained for numerous other quasars. In fact, to be visible at such great distances, the luminosity of some quasars must be greater than the combined output of a hundred galaxies!

Rapid changes in the energy output of some quasars, such as 3C454.3, over a few weeks indicate that quasars are very small. If the quasar were large, the change in light output would be gradual. This is because it would take longer periods of time for the light, which has a finite speed, to travel across the entire quasar. For example, the light-emitting portion of quasar 3C273 is no larger than our Solar System! How can an object only a few light hours in diameter produce the energy of more than a billion suns?

Most astronomers believe that quasars are very active galaxies, closely related to Seyfert galaxies, that have massive black holes at their centers. In 1982, scientists at the California Institute of Technology detected a faint glow of starlight surrounding quasars 3C48 and 3C273 and discovered the spectra of hot, young stars. The distribution of starlight matched that found in elliptical galaxies. In addition, the temperatures found in quasar gases seemed to correspond to the conditions predicted by the presence of a central black hole—a billion solar masses 5 light hours in diameter! Similar black holes have been predicted for radio galaxies. Radio images showing jets of ejected matter indicate that the black holes may be rapidly spinning, and have been named **spinars**.

When we observe a distant quasar, we are actually looking back in time, as the light has been traveling at a rate of 6 trillion miles a year (the speed of light) since its creation. Since quasars have been detected as far away as 10 billion light-years, they probably formed early in the history of the universe—no more than a few billion years after the initial Big Bang. Because no quasars have been found nearby, some astronomers believe that quasars are galaxy-like objects with an extremely bright nucleus. No doubt quasars existed in the distant past under very different conditions.

Some astronomers, such as Halton Arp

at the Mount Wilson Obsevatory, believe in another theory. Arp contends that quasars are not as distant as they appear. If quasars are closer, then they are also not as energetic or bright, so the mystery of their energy production would be partly answered. As evidence, he claims to have found numerous quasars in association with nearby galaxies. But the mechanism for their extreme red-shifts remains unexplained and the debate continues.

The Evolution and Shape of Galaxies

In the very early universe, when there were no galaxies or stars, matter was evenly distributed in a vast cosmic cloud. Now and again, atoms would stick together and form a temporary clump of matter. If the condensations were large enough and gravity held the atoms together, the clump grew larger. It remained distinct from the surrounding gases that filled the universe. If it grew large enough, it was the beginning of a galaxy, or protogalaxy. (This is basically the same condensation process describing the formation of the Sun and planets in Chapter 8.)

A protogalaxy developed before any stars formed in the cloud. As the protogalaxy continued to contract, its density increased and matter was gravitationally drawn toward the center. Pockets of gas and dust swirled, condensed, dissolved, and recondensed throughout the cloud. Whenever a small pocket was dense enough, it became a protostar and followed along the Main Sequence. Eventually, most of the gas and dust formed the stars with very little remaining throughout the galaxy. (All the galaxies in the universe are still alive since it will take trillions of years—far greater than the present age of the universe—for the longest living stars to become black dwarfs.)

At one time, it was believed that the Hubble classification of galaxies represented the evolutionary stages of a typical galaxy. Today, astronomers are convinced that the galaxies formed all at once when the universe was about one billion years old. Their differences in shape are largely attributed to the local conditions under which they formed.

Although there are no definitive answers to why there are different shapes of galaxies, some plausible explanations exist. One of the most widely accepted theories is that the shape of a galaxy can be determined by the distribution and density of the material present in the protogalaxy and the amount of angular momentum, or spin, that the original gaseous cloud experienced upon formation. The more spherical galaxies spun very slowly so they did not flatten out as they condensed. Elliptical galaxies were spinning at a more rapid rate, and therefore exhibit a greater degree of flattening. Spiral galaxies had the greatest degree of initial spin and therefore are the most disk-like of the three. In addition, the nucleus of a spiral galaxy contained the bulk of the original material. And since it was not as affected by rotational forces, it shows the same general shape of the more spherical galaxies as well as the same population of stars.

There is a problem in this explanation: What causes the spiral arms to form? If the Sun has made 20 rotations around the galactic center since it was formed, why haven't the spiral arms twisted up tightly around the center? How do the spiral arms maintain their shape? Until recently, there was no satisfactory answer. But now, observations made by radio telescopes indicate that our galaxy's nucleus (and other spiral galaxies) continually eject hydrogen gas, contributing to the raw material needed for new star formation. According to the most recent popular theory, called **wave density**, small variations in the gravitational field of the galaxy change the rotation speed of material around certain parts of the galactic center. Any decrease in speed allows material to accumulate and leads to the formation of new stars. As new stars enter the main sequence, they create areas of greater luminosity. These brighter areas appear to the observer as the familiar spiral patterns.

A good analogy is the flow of traffic on a highway. If for any reason some cars reduce their speed (to rubberneck at an accident, for example), cars pack together and drivers are caught in a traffic jam. The cars do not even have to stop. Even after an accident has been cleared, the annoying bunching up of vehicles persists. Therefore, spiral arms may be considered cosmic traffic jams!

Large-Scale Structure in the Universe

As larger portions of the sky were mapped by sophisticated radio telescopes, astronomers became aware that the universe tended to organize itself into larger structures. Beginning in the early 1930s, Fritz Zwicky observed that just as cities join states and states form nations, galaxies form clusters and clusters form superclusters—all joined by mutual gravitational attraction.

On the small scale, there are binary pairs of galaxies orbiting some common center, like the Whirlpool Galaxy (M51, or NGC 5194) and its companion, NGC 5195, both some 15 million light-years away. There are also clusters of only one or two dozen galaxies, such as the **Local Group** (to which our own Milky Way belongs). The Local Group is a cluster of galaxies bound together by gravity and all travelling toward the same point in space. Included in the local group are the Large and Small Magellanic Clouds, M31, M32, NGC 205, and M33. In the late 1960s, two new members of the local group were discovered using observations in the infrared and at radio wavelengths. Now called Maffei I and II, after the Italian astronomer who first detected them, they appear to be an elliptical and a spiral galaxy, respectively. They may, however, prove to actually be members of another, relatively nearby cluster of galaxies. At best, they are on the outskirts of the Local Group and may be only passing through.

Some clusters of galaxies contain many hundreds or even thousands of individual galaxies. The nearest cluster appears in the constellation Virgo and is called the Virgo cluster. It contains many thousands of galaxies, covering nearly 6° of the sky—roughly equal to the diameter of 12 full Moons—and is about 60 to 75 million light-years away. Another prominent cluster is the Coma cluster, of the constellation Coma Berenices. The Coma cluster is nearly spherical in shape. It resembles a globular cluster but with a concentration of galaxies, rather than stars, toward the center. The Coma cluster is roughly 300 million light-years from us. One of the more distant of the giant clusters is the

Hercules cluster some 500 million light-years away.

The Intergalactic Medium

What is in the space between the galaxies? If all the galaxies formed from huge clouds of gas and dust, it is safe to assume that only a fraction of this material condensed into visible stars and nebulas, leaving intergalactic space filled with some form of non-luminous matter. (The implications for the ultimate fate of the universe regarding this dark matter will be examined in the next chapter.) Observations made by the Einstein X-Ray Observatory in the mid-1980s, indicated that hot, intergalactic matter, with temperatures greater than 10^6 °K, exists in localized regions. To what extent this material pervades the universe has not yet been determined.

In addition, other X-ray observations made of the Perseus cluster of galaxies has shown the presence of iron atoms in the intergalactic medium. If the hot gas of the intergalactic medium is primordial hydrogen and helium created at the birth of the universe—and was never part of any galaxy where only the nuclear reactions of stars could form iron—how did the iron atoms get from interstellar space to intergalactic space? From X-ray observations made by the Einstein X-Ray Observatory of M86, a galaxy in the Virgo cluster, a long plume of matter was detected trailing from the rear of the galaxy. M86 is currently speeding through the center of the Virgo cluster at 3 million miles per hour. Scientists suspect that the intergalactic medium is resisting this motion and is stripping the galaxy of iron and other heavy elements. These elements then mix with the intergalactic medium in a process known as **ram pressure stripping**.

Gravitational Lenses

In 1915, Albert Einstein published a theory that introduced the idea of **gravitational lensing**. This theory states that the gravity of a massive object would bend any starlight that passed within its field. Four years later, in order to prove Einstein's idea, an international team of astronomers made accurate measurements of the effect of the Sun's gravity on the light from the stars in the Hyades star cluster. They made their observations during a solar eclipse in order to photograph the stars as their light skimmed the solar disk. Einstein was discouraged by the difficulty of detecting this lensing effect—especially when the two stars were so distant that their chances of perfect alignment were small.

One year later, Fritz Zwicky suggested that the chances of detecting gravitational lensing would be far greater if the source and the lens happened to be massive galaxies rather than individual stars. Unfortunately, his paper received very little attention.

In the early 1990s, astronomers at the Kitt Peak Observatory in Arizona observed a very odd pair of quasars near the Big Dipper. They discovered that the pair was actually a double image of the same single quasar. The quasar's light had been split by the gravity of a massive (but very faint) galaxy that lies in a direct line of sight between the Earth and the quasar. To date, six credible examples of gravitational lensing have been detected. With the discovery of more candidates, astronomers will have a better understanding how light bends because of gravity—and of the distribution of mass in the universe.

Superclusters

On a very large scale, galaxies seem to congregate and form clusters of clusters—or superclusters. A recent study of about one million galaxies showed that the largest structures in the universe may be walls and strings of galaxies that stretch for several hundred million light-years.

The Milky Way galaxy and the other members of the Local Group, along with about 50 other clusters, are part of the Local Supercluster. The Virgo cluster some 60 million light-years away, forms the center of the Local Supercluster where the population of galaxies is about 10 to 100 times more dense than in the region of the Local Group.

Voids

Viewed on the same scale, the spaces separating superclusters seem surprisingly empty of any matter. Dubbed **voids**, they are hundreds of millions of light-years across but contain no galaxies. To compare, more than one billion Milky Ways could fit into such large volumes of space. Astronomers believe that superclusters take up around 5 percent of the volume of the universe and the rest is empty space. However, a more popular theory proposes that vast galaxies of dark matter fill the empty spaces.

CHAPTER THIRTEEN

The Origin and Fate of the Universe

KEY TERMS FOR THIS CHAPTER

big bang
cosmic background radiation
cosmology
entropy
horizon distance
Hubble constant
inflationary epoch
megaparsec
oscillating
steady-state
velocity-distance relationship (Hubble's law)

We began our discussion of astronomy by explaining humanity's unceasing fascination with the objects in the sky. Although we have made many discoveries about our universe, the solutions to such eternal questions as how, when, and by what means the universe originated still elude us. The branch of astronomy that is specifically concerned with these difficult questions of the universe's origin and fate is known as **cosmology**.

Through observations of the physical laws of the universe, scientists have determined with some degree of certainty that our universe had a beginning. They have reconstructed a reasonable life history of the universe—from the fraction of a second after its birth to the present era. If the universe had a beginning, then logic demands that it must also have an end (which will occur many billions of years in the future). The discovery that hydrogen—the basic fuel of creation—is running out in the universe, seems to confirm this idea. Just what type of ending awaits the universe is currently a hotly debated topic in cosmology. Astronomers hope to discover the universe's fate by finding and understanding the nature of the so-called hidden (or dark) matter believed to make up the bulk of the universe.

The Expanding Universe

In a series of studies begun in 1912, astronomer Vesto Melvin Slipher recorded the spectral red-shift of 15 nearby spiral galaxies. The red-shifts of each galaxy indicated they were moving away from the Earth at high speeds—some in excess of two million miles per hour! Why should these galaxies be receding from us? The motions in the universe should be random, with some galaxies moving away from us and some moving toward us. Yet his studies led to the conclusion that the Earth occupied a special position in the universe—dead center.

In the 1920s and 1930s, Edwin Hubble and Milton Humason used the 100-inch telescope at Mount Wilson to measure the spectra of many more faint galaxies. (Humason was a remarkable man. He had only an eighth-grade education and worked his way up from mule driver to janitor and then to Hubble's assistant.) Their studies confirmed Slipher's early findings—that no matter where one looked, every galaxy was moving away from the Earth. Hubble also discovered that the more distant the galaxy, the faster it was receding. In other words, if a galaxy 10 million light-years away were receding at a velocity of 100 miles per second, then a galaxy at 1 billion light-years distance would be moving away at a speed of 10,000 miles per second. This phenomenon is now known as the **velocity-distance relationship**, or **Hubble's law**. Further studies using the 200-inch telescope at Mount Palomar after World War II confirmed that the universe was expanding with the Earth at its center.

Later studies showed that the Earth only appeared to be in the center. Analogy explains the true nature of our universe's expansion. Imagine a cake full of raisins baking in an oven. Each raisin represents a cluster of galaxies with an observer located on each raisin. As the yeast in the mix causes the cake to swell, the observer sees every raisin-galaxy move away from every other raisin-galaxy. Each observer on any another raisin-galaxy would have the same impression. If the cake were infinitely large (so that an observer could not see the edges from the interior), the observer, no matter what the location, would not be able to find the center. This is most likely the true nature of our universe (Figure 13.1).

The Big Bang

Since the universe is expanding and all of the galaxies are moving apart, scientists assume that at one time they were closer together. Turning the cosmic clock back far enough, we can imagine a tightly packed dense ball under extreme temperatures and pressures. The brilliant explosion of the mass sent matter flying outward in all directions—marking the birth of our universe. This **big bang** cosmology is the one most widely accepted today.

When did our universe flash into existence? In order to know the universe's birthdate, it is necessary to calculate exactly how fast the galaxies are receding from one another. The uniform proportion of a galaxy's recessional speed to its distance is known as the **Hubble constant**—abbreviated H, and often expressed in kilometers per second per **megaparsec**, or 3.25 million light-years. If this number is very great, then the galaxies are rapidly

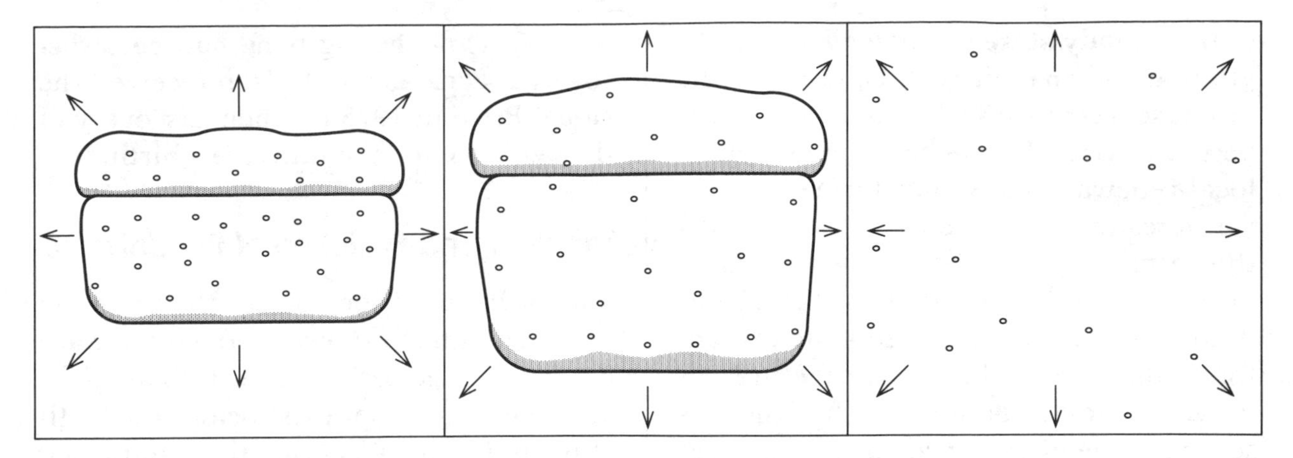

Fig. 13.1. The expansion of the universe compared to the expansion of raisins in a baking cake.

moving away and were packed together a relatively short time ago. This would make the universe astronomically young. If the Hubble constant is less, then the galaxies are moving slowly and time has allowed for greater distances between the galaxies. Thus, the universe would be much older.

Unfortunately, scientists do not agree on the exact value for the Hubble constant. As the distances to remote galaxies are measured more accurately, the value for the Hubble constant, and thus the age of the universe, changes. Today, most scientists estimate the age of the universe to be somewhere between 10 billion years (for H = 100 kilometers per second per megaparsec) and 20 billion years (for H = 50 kilometers per second per megaparsec), with 15 to 18 billion years an accepted average. As astronomical instruments take more accurate measurements, astronomers will gain a better understanding of the Hubble constant and therefore the age of the universe. For example, one of the primary objectives of the Hubble Space Telescope (named after Edwin Hubble) is to measure the distances to remote galaxies more accurately.

An additional complication arises when computing the exact age of the universe. We know that the universal force of gravity is the cosmic glue that binds everything together. Therefore, gravity must be tugging all matter in the universe, and slowing down its rate of expansion. (This is similar to how a stone tossed in the air loses its speed as the Earth's gravity halts its upward flight.) The slowing down of the universe's expansion depends on the total mass contained within the universe—a topic that will be explored later in this chapter.

The Steady-State Theory

Not all scientists have embraced the big bang cosmology. In 1948, British astronomer Fred Hoyle (1915–) and several colleagues proposed a different explanation for the existence of the universe based on the red shifts of distant galaxies. In this theory, the universe is constantly renewing itself by continuously creating hydrogen to replace the empty spaces left by the retreating galaxies—without the need for a cosmic explosion. In effect, proponents of this **steady-state** cosmology deny a beginning and an end. They believe that the universe has always been and will always be there, virtually unchanging except for local renewals like stellar formation.

Interestingly, it was Hoyle who coined the term "big bang" to show his disapproval for his theory's greatest rival. Later in his career, while attempting to explain the abundance of helium in the present universe, Hoyle adopted much of the reasoning of the big bang cosmology, contradicting many of his earlier conclusions. The steady-state cosmology was soon put aside—especially with the discovery of the echo of the explosion that created our universe.

Cosmic Background Radiation

If the big bang actually happened, then the exploding universe must have been filled with intense radiation in those first few moments. The strength of this primordial fireball would have grown weaker as the universe continued to expand and cool. But a small remnant should exist today, detectable by sensitive radio telescopes.

In 1965, Arno Penzias (1933–) and Robert Wilson (1936–) of the AT&T Bell Laboratories in Holmdel, New Jersey, were using a 20-foot antenna to search for sources of radio noises from the Milky Way. They detected an annoying hiss from the receiver that they could not eliminate. At a characteristic background temperature of about 3° K, Penzias and Wilson made the most important cosmological discovery in the latter twentieth century: **cosmic background radiation**. With some theoretical support from Princeton University physicists led by Robert H. Dicke, the echo (now a whisper) of the big bang had been discovered. Penzias and Wilson received the Nobel Prize in 1978 for their discovery of the evidence for the universe's birth.

The Birth and Evolution of the Universe

The exact conditions before the big bang are impossible to know, although there are numerous theories. Time, as well as space and matter, did not exist before the birth of the universe. Accepting this limitation, scientists can describe the major events in the evolution of the universe with some confidence, starting from time zero: the big bang.

According to the standard big bang cosmology, the universe was a fireball, infinitely hot and dense the first moment after the big bang. (Astronomers believe that the closest comparison is the singularity of a black hole.) It enclosed a space approximately equal to the Earth's orbit around the Sun, with temperatures as high as 100 trillion degrees. All the matter and energy in today's universe was contained in the rapidly expanding fireball. The temperature and radiation were so intense that every type of particle smashed into each other, being destroyed as soon as they were created—reprocessed over and over again.

Within several minutes the temperature had dropped below 1 billion degrees and nuclear reactions slowed down. Matter become more stable with the formation of protons, neutrons, electrons, and photons. The young universe was filled with these energetic particles swimming in a bath of hot radiation. The only other nuclei that were created, apart from the hydrogen nucleus (a single proton), were those of helium (2 protons and 2 neutrons). In fact, all the matter in the universe consisted of about 75 percent hydrogen and 25 percent helium, formed in the first 30 minutes of the universe. According to the big bang theory, the formation of all the heavier elements did not begin until a billion years later, when stars began to form.

For the next several hundred thousand years, the ball was packed with electrons and protons. In fact, the density was so great that as soon as a photon was emitted, it was reabsorbed and scattered. No light could escape and the universe was dark. When the temperature of the expanding universe cooled to about 3,000° K, the collisions between the energetic particles slowed enough to allow protons to capture electrons and form more stable hydrogen and helium atoms. Matter and radiation successfully separated, or decoupled, and the universe became transparent, shining with a reddish light as bright as the surface of the Sun. The fireball was now about $1/_{1,000}$ the size of the present universe and continued to expand.

When the universe was between several hundred million and a billion years old, the gaseous clouds of primordial hydrogen and helium cooled and eventually condensed into protogalaxies. Soon after, the first generation of stars began to form and the synthesis of the heavier elements continued. In the following 15 billion years, the universe continually expanded, stars formed and died—producing the universe of today.

A Problem with the Big Bang

Although the theoretical predictions of the big bang cosmology seem to agreee with actual observations of the universe, a major question about the uniformity of the 3° K background radiation remains a problem. How could a perfectly smooth and homogenous temperature result from the chaos of the rapidly expanding fireball? Why aren't there random differences in the temperature? Yet careful measurement has shown that the temperature of the remnant fireball is exactly the same in all parts of the universe.

The problem arises in the earliest microsecond of the universe's birth. The finite speed of light (186,000 miles per second) limits how fast information about any physical process can travel from one point in space to another. In other words, object A cannot affect object B faster than the speed of light. From the instant of the big bang, there has been a limiting distance across space called the **horizon distance**—the distance a light signal could have traveled since time zero. When the big bang began, sources of energy expanding in opposite directions were already further apart from each other than the horizon distance. How could they have the exact same temperature—especially since they were so far apart? Proponents of the standard big bang theory avoid the horizon problem by assuming that the conditions of uniformity were present from time zero. Therefore, the universe would evolve uniformly as a natural consequence of the known physical laws.

One hypothetical answer to the horizon problem was proposed by Alan Guth of Cornell University in 1979. His theory, known as the **inflationary epoch**, describes possible events between time zero plus 10^{-39} second and plus 10^{-30} second—the very beginning of creation. According to one possible scenario, our universe popped into being from something called the quantum vacuum. It was an infinitesimal bubble of space only one-trillionth the size of a proton that had the good fortune not to disappear back into the vacuum. Other tiny bubble universes also appeared in the same way, all inside a much larger bubble. At a certain point, our universe released a tremendous amount of energy and underwent a violent inflation or growth. Because a greater horizon distance already existed (the bubble that contained our universe), no part of our bubble could be hotter or colder—making the temperature of the universe homogeneous no matter how it expanded.

The Fate of the Universe

When astronomers speak of the size of our universe, they are referring to the observable universe; that is, all the matter that emits energy in the form of light or radio waves. Every particle of mass in the universe acts on every other particle to tie the universe together. This universal gravity must be slowing down the expansion. Will the universe coast outward forever (a flat universe)? Will the rate of expansion become more rapid (an open universe)? Or, will gravity be sufficient enough to halt the outward expansion of the universe and bring all the galaxies back together (a closed universe)?

The answer may lie in how much mass the universe actually contains. The ratio of the density of the universe to the critical density is labelled by the Greek letter omega (ω). If $\omega = 1$, the universe is flat; if $\omega < 1$, the universe is open; and if $\omega > 1$, then the universe is closed. The most recent mass measurements of all the visible stars in the galaxies fall far below the critical mass—less than 10 percent. Does this mean that the universe is open? Astronomers are not ready to say yes because there is evidence that not all the matter in the universe has been detected.

The Missing Mass

Astronomers had long believed that they could locate all the mass in the universe by studying the distribution of starlight. Fifty years ago, Fritz Zwicky and Sinclair Smith at the California Institute of Technology studied the motions of clusters of galaxies and concluded that the gravitational pull of some nonluminous masses kept the clusters from flying apart. Other astronomers (most notably Vera Rubin and her colleagues at the Carnegie Institution in 1977) engaged in a series of studies that confirmed the existence of some hidden mass surrounding galaxies. This mass, in the form of gigantic halos (see Chapter 12), helps determine a galaxy's observed orbital motions. In other words, if the invisible mass did not exist, the gravitational effect of a galaxy's luminous mass would not be enough to explain the galaxy's rotation. At the present time, astronomers are busy searching for more evidence of dark matter.

What this dark matter is remains purely hypothetical. There are candidates from the field of particle physics (the study of subatomic particles) such as the neutrino, photino, axion, and gravitino. Another theory calls for the existence of numerous baby black holes to account for the hidden mass in the universe. Obviously, much

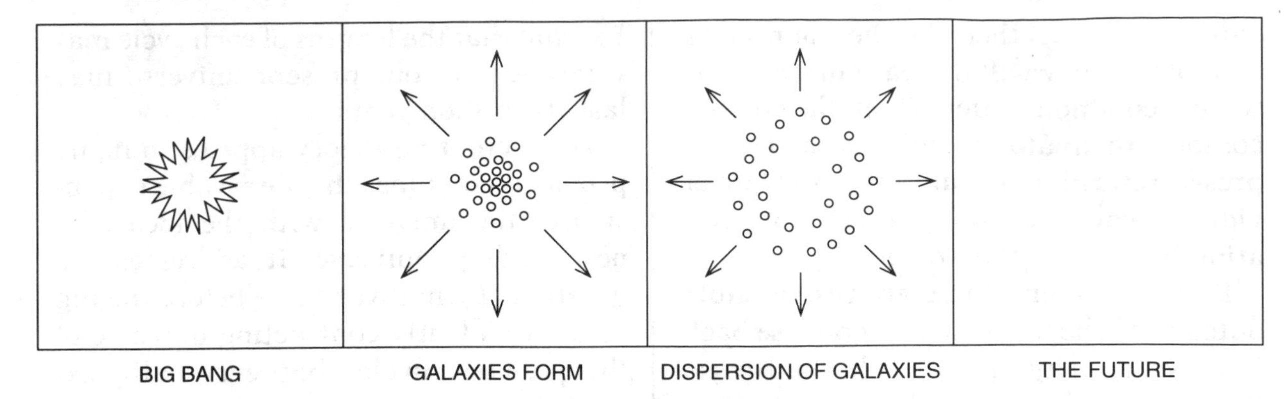

Fig. 13.2. The future of the universe according to the standard big bang cosmology.

more research needs to be conducted before scientists have the answer.

If there is not enough mass to halt its expansion, then the universe will have a cold, dark ending. As all the hydrogen in the galaxies is used up, fewer new stars will form. And as galaxies and stars travel outward, the empty spaces between the galaxies will grow larger. The universe, once ablaze with light, will then fade into darkness. It will have reached maximum **entropy**—the final state of disorder of any organized system (Figure 13.2).

The Oscillating Universe

A totally different fate may await the universe if there is enough mass to halt the expansion. In **oscillating** cosmology, the gravitational effect of the universe's mass is sufficient to eventually stop the movement of all the galaxies. At one point in time, the galaxies will stand poised between the outward momentum created by the explosive primordial fireball and the inward tug of gravity. Then the galaxies will reverse their direction and begin to

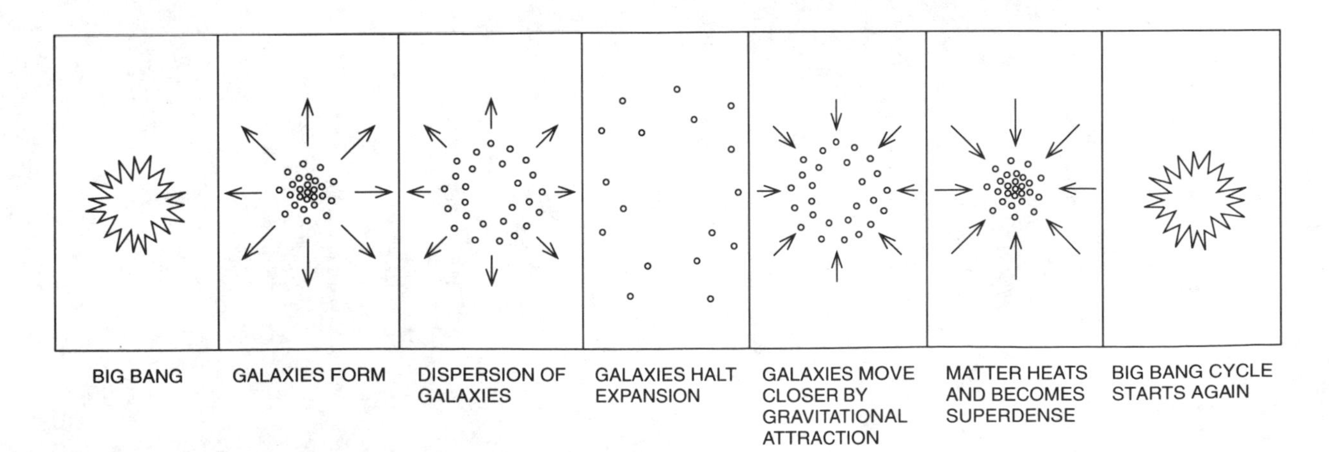

Fig. 13.3. The future of the universe according to the oscillating cosmology.

move closer together. As they approach each other, they will increase in speed toward a common center. When they finally contact, their atoms will mix and compress—returning the universe to a state of violent heat and chaos similar to its creation billions of years before.

From this point, there are two possible outcomes: The universe will collapse back into a singularity—infinitely hot and dense (the big crunch), or the collapsing universe will rebound and develop into a new fireball, eventually expanding, cooling, and forming new galaxies and stars. The process then repeats itself again and again, with each successive cycle creating new and different universes (Figure 13.3). John Wheeler, formerly of Princeton University, has said that the lengths of each cycle may vary and that our present universe may last 100 billion years.

The oscillating theory appeals to many people. It combines the observable expansion of the universe with the idea of a never-ending universe. It addresses the question of what was there before the big bang: A violently contracting universe of the previous cycle that eventually exploded to form ours. No one really knows how the universe began. In fact, some astronomers believe that the answer may prove to be a combination of the standard big bang and oscillating theories. But for now, we must wait until scientists discover new clues to help solve the mysteries of the beginning and end of our universe.

CHAPTER FOURTEEN

Observing for Amateurs

KEY TERMS FOR THIS CHAPTER

asterism
celestial sphere
circumpolar stars
local celestial meridian

The Basics of Observing

Amateur astronomers enjoy studying the night sky with the naked eye, binoculars, or a small or medium-size telescope. It is preferable to be as far away from city lights as possible because the darker the sky, the more celestial objects are observed. In addition, a wide and unobstructed view of the sky at an observing site allows for uninterrupted tracking of celestial objects.

This chapter introduces the basic information needed for exploring the night sky. Monthly star charts are included with descriptions of some of the more interesting celestial features. (The star charts are for observers living in mid-northern latitudes.)

The Constellations

Brighter stars in the dome of the sky, or **celestial sphere**, are classified into groups known as constellations. As long ago as 150 A.D., the Greek astronomer Ptolemy listed the names of forty-eight constellations in a catalogue called the *Almagest*. In addition to the ancient Greeks, various other cultures, such as the Chinese and Native Americans, imagined that groupings of bright stars formed pictures of gods and goddesses, heroes and heroines, animals, or objects. During the seventeenth century, European astronomers added more constellations, such as Reticulum the Reticle, Telescopium the Telescope, and Sextans the Sextant, to honor the scientific instruments that helped them map the sky.

Modern astronomy recognizes 88 constellations, each with its own clearly defined boundaries and original name. The abbreviations of the names of the constellations are those of the International Astronomical Union (IAU). (See Appendix B.) The 88 constellations completely cover the celestial sphere.

The Magnitude of Stars

In the second century A.D., the ancient Greek astronomer Hipparchus classified stars into six classes according to their apparent brightness, or magnitude. He arbitrarily designated the 20 brightest stars as the 1st magnitude; the next 50 stars as the 2nd magnitude. Finally, Hipparchus gave the designation of 6th magnitude to several hundred stars that were barely visible to the unaided eye (Figure 14.1). Hipparchus was measuring the apparent magnitude of the stars. Some stars are actually brighter than others, but appear fainter because of their greater distances. (The true brightness of a star is known as its absolute magnitude. Astronomers measure the absolute magnitude by determining a star's brightness as if it were located 10 parsecs [32.5 light-years] from the Earth [see Chapter 10]).

In the nineteenth century, more precise measurements of stars demanded a new method of classifying a star's magnitude. Astronomers then introduced decimal di-

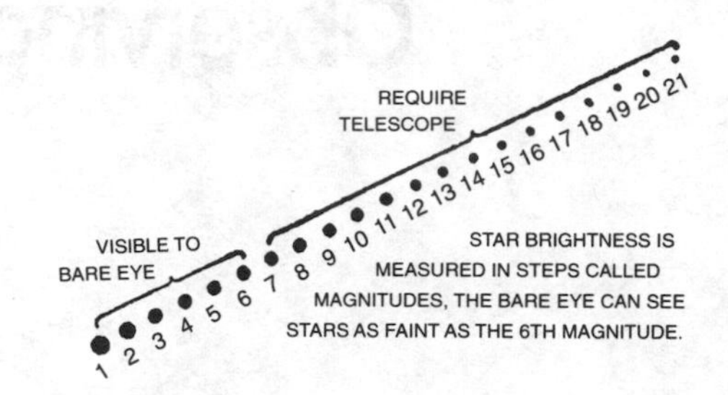

Fig. 14.1. The relationship between brightness and magnitude.

visions for each order of magnitude. Thus, a star at magnitude 5.5 has an apparent brightness between magnitude 5.0 and 6.0. For example, the North Star, or Polaris, with a magnitude of 2.1, is slightly fainter than a star of magnitude 2.0.

A simple relationship exists between the magnitudes of stars. For example, stars that are magnitude 2.0 appear 2.5 times (more precisely, 2.512) brighter than stars that are magnitude 3.0; stars that are magnitude 3.0 appear 2.512 times brighter than stars of magnitude 4.0, and so on. A 1st magnitude star is one million times brighter than a 15th magnitude star!

Hipparchus made several mistakes in his original system of classification. The 20 stars he designated as 1st magnitude were eventually reclassified. In addition, astronomers introduced magnitudes of less than one, zero, and negative values. For example, the brightest star in the sky for northern observers is Sirius in the constellation Canis Major; it has an apparent magnitude of −1.48. On the same scale, the magnitude of Arcturus (α Boötes) is exactly 0.0; while the magnitude of β Centauri is 0.6. At the extreme end of the scale is our Sun, with a magnitude of −26.7! Our Sun's absolute magnitude is +4.0, but because it is so close to the Earth, its apparent magnitude is great.

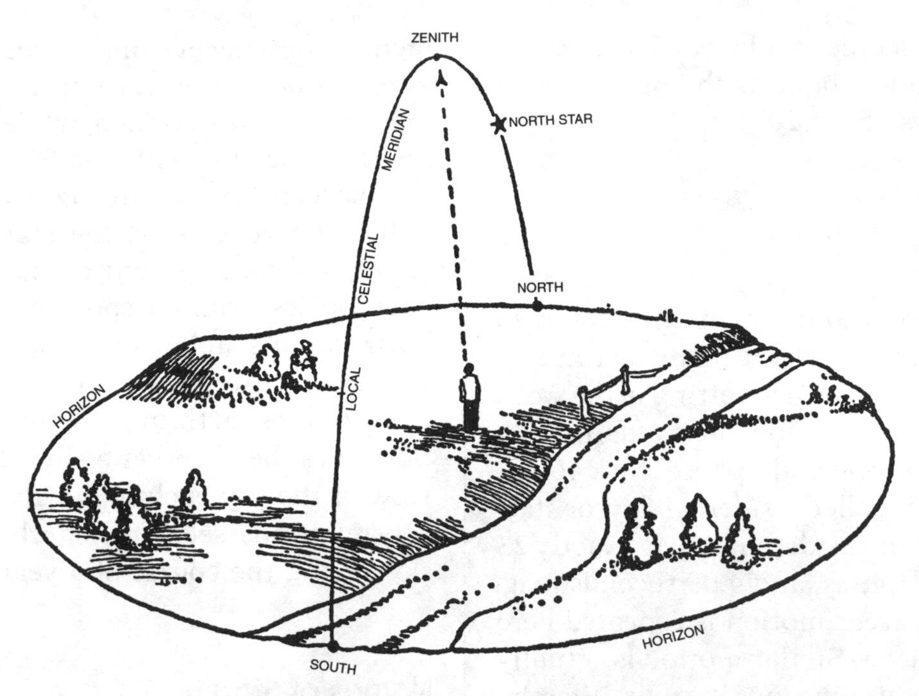

Fig. 14.2. The local celestial meridian is an imaginary circle in the sky. It goes through the north and south points on the horizon and the zenith. Observe the stars at about 8 P.M. near the middle of the month and in a year's time you will see the entire parade of stars pass this line.

Measuring Distances in the Sky

The location and/or distance between any two observed objects in the sky is most often measured in degrees. For example, a complete circuit around the horizon is 360°. A line drawn from the exact northern point on the horizon, through the point directly overhead (the zenith) and down to the exact southern point on the opposite horizon measures 180°. This is known as the **local celestial meridian** (Figure 14.2).

Other distances measured by the observer are much smaller. For example, the diameter, or angular distance, of a full Moon is about one-half a degree (Figure 14.3). The angular distance between the two pointer stars, Dubhe and Merak, in the bowl of the Big Dipper, is about five degrees (Figure 14.4). Ten full Moons placed side by side would equal the distance between these two stars.

For the amateur astronomer, a convenient way of measuring average angular distances requires only the observer's outstretched hand. With the arm fully extended, note the width of your index finger. This width covers an angular distance of

Fig. 14.3. The angular distance of the full Moon is about half a degree.

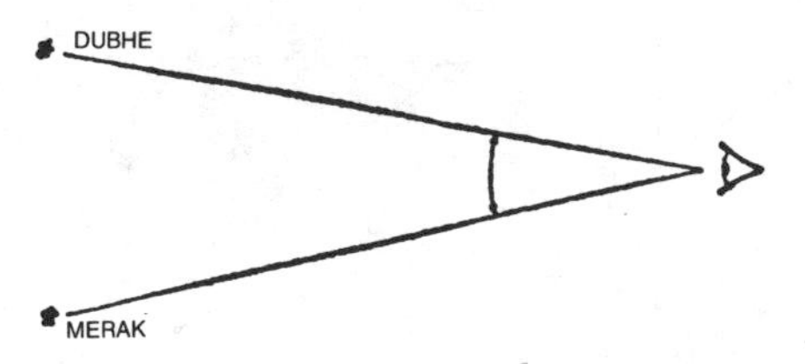

Fig. 14.4. The angle formed by Dubhe and Merak at the eye of a terrestrial observer is close to 5°.

about one degree; a clenched fist covers about 10°; and a hand with fingers spread wide, covers about 20° (Figure 14.5).

The Apparent Motion of Stars

The stars seem to follow paths similar to the Sun and Moon, rising in the east, arcing across the sky, and then setting in the west. A complete revolution of a star from a point on the celestial sphere back to the same point is called a sidereal day, or star-day. On Earth, a sidereal day is exactly 23 hours, 56 minutes, and 4.09 seconds long. A star's apparent motion is repeated here for good reason: Stellar motion is actually the Earth spinning on its axis, thereby causing the stars to seem to move across the sky.

The fact that stars complete one revolution in less than 24 hours is significant for the observer. It means, of course, that the stars make just over one revolution in a 24-hour period. The difference between a sidereal and a solar (24 hour) day is 3 minutes and 56 seconds. Therefore, a star that appears on the eastern horizon at eight o'clock on a Sunday evening will be slightly above the horizon at eight o'clock on the following evening. At the same time a month later, the star will be noticeably higher above the horizon; while after three months, the star will be a 90° away from the eastern horizon at eight o'clock. Finally, one year later, the star will have completed an apparent circle, ending up back on its original spot on the eastern horizon. This stellar movement is also an apparent movement. The actual movement of the Earth in its orbit around the Sun gives the observer a slightly different view of the sky each night. It is also why constellations seem to march across the sky during the course of a year.

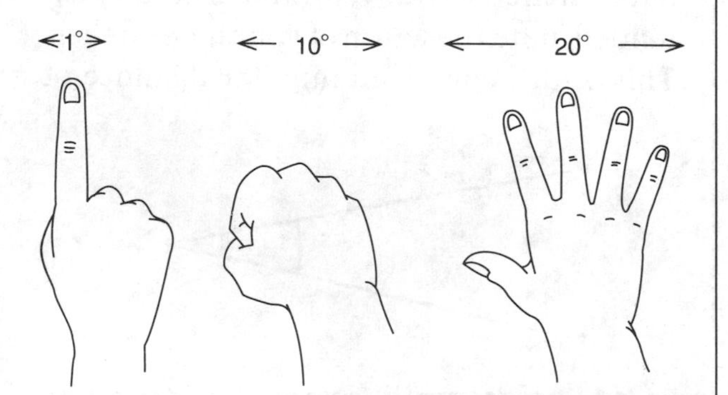

Fig. 14.5. Measuring angular distances with the aid of the hand.

Names of Stars

Generally, a star is identified by a Greek letter and the name of the constellation in which the star appears. For example, the stars in Ursa Major (the Great Bear), where the Big Dipper is the most familiar pattern, or **asterism**, are known as α (alpha) Ursae Majoris, β (beta) Ursae Majoris, and so on.

Historically, the brightest star received the α (alpha) designation, the second brightest received the β (beta) designation, and so on. There are, however, several exceptions to this rule. For example, in the constellation Gemini (the Twins), the brightest star is the beta star and the next brightest star is the alpha star. The third brightest star in Gemini is γ (gamma) Geminorum.

If the number of bright stars in a constellation exceeds the number of Greek letters, numerals are used, such as 16 and 17 Draconis—magnitude 5.4 and 5.5 stars that form a binary star system in the constellation Draco the Dragon. In addition, upper-case Roman letters are also used for variable stars, as was discussed in Chapter 10. Lastly, some 60 well-known stars are often referred to by their proper names. Among these are Polaris, or the North Star (α Ursae Minoris), Sirius (α Canis Majoris), and Regulus (α Leonis).

In addition to naming and numbering the stars, many of the brighter clusters, nebulae, and galaxies have been catalogued and named. One such listing was compiled by Charles Messier (1730–1817) in the eighteenth century, with each celestial object known by its Messier, or M numbers. For example, M1 is the Crab Nebula in Taurus (it is also the first entry in Messier's catalogue). William Herschel (1738–1822) completed a survey of the sky in the 1780s and published a catalogue of his findings. About a century later, John L.E. Dreyer (1852–1926) published his *New General Catalogue of Nebulae and Clusters* (*NGC*). Most of the Messier objects appear in the NGC. For example, M1 is also identified as NGC 1952. Whereas Messier compiled a list of just over one hundred objects, Dreyer was able to catalogue thousands of celestial objects due to greatly improved instruments and observing techniques.

How to Use a Star Chart

When observing the night sky, beginners often have difficulty using a star chart because celestial directions are more confusing than terrestrial maps. Remember that an observer is on the outer surface of a sphere (the Earth) observing the stars on the inner surface of another sphere (the celestial sphere).

To use the charts properly, hold each chart vertically so that the horizon direction on the map is aligned with your actual terrestrial horizon. For example, if you wish to view the southern portion of the sky, hold the sky chart with its southern horizon aligned with your southern horizon. For a different portion of the sky, the chart must again match your chosen horizon. Only then will the map's stars match those in the sky. With practice this procedure will become almost automatic. However, in order to simplify instructions, the following examples of observations will be made with the observer facing the southern horizon.

The Night Sky Each Month

January

As the year begins, the brilliant star-filled band of the winter Milky Way stretches from the constellation Canis Major near the southeastern horizon, through the zenith near Taurus and Auriga, then disappears on the northwestern horizon near the constellations Cepheus and Cygnus. In addition, some of the brightest stars are in the winter sky. For example, Sirius is the brightest star in the northern hemisphere and is relatively close (8.6 light-years from Earth). It is also called the Dog Star because it is in Canis Major, the Great Dog. Observers using binoculars or a small telescope can see the open star cluster M41 (NGC 2287) about 4° south of Sirius. At magnitude 6.0, it is sometimes visible to the naked eye under excellent viewing conditions. It is a fairly rich cluster that measures about ½° in diameter and contains approximately 25 bright stars and many fainter ones.

Sirius is one point of an asterism known as the Winter Triangle. The other two points include the red supergiant Betelgeuse in Orion (to the northwest or upper right), and Procyon in Canis Minor (to the northeast or upper left). In February, conditions for viewing the Winter Triangle are a little better as it is then higher above the haze of the horizon.

Almost due south is the constellation Orion the Hunter. One of the most easily recognizable asterisms in Orion is his belt, formed by a line of three evenly spaced stars. Just below the belt is part of Orion's sword, M42 (NGC 1976)—the Great Orion Nebula. The Orion Nebula is about 1,500 light years away and appears as a fuzzy star to the naked eye. Under moderate magnification, this object appears as a magnificent luminous cloud surrounding the star θ^1 (theta^1) Ori. Small telescopes can resolve θ^1 Ori into four individual stars known as the Trapezium. The four stars (magnitudes 5.1, 6.7, 6.7, and 7.9) are very hot and illuminate the nebula.

Two other bright stars are located in Orion. One is Betelgeuse, or α (alpha) Ori—a supergiant whose reddish color is apparent to the naked eye. (Astronomers believe Betelgeuse may become a supernova in the near future.) At a distance of 310 light years, it marks the right shoulder of Orion. The other bright star is Rigel, or β (beta) Ori, marking Orion's left leg. Rigel is the 7th brightest star in the sky (magnitude 0.1) and is a bright blue-white color.

To the right of the northwest corner of Orion's border is a V-shaped asterism known as the Hyades. This loose, open cluster of one-billion-year-old stars spans over 5° and forms the face of the constellation Taurus the Bull. The stars are approximately 140 light-years away and are white in color, contrasting with the magnitude 0.9 orange-red star Aldebaran, or α (alpha) Tauri, which marks the right eye of the bull. Aldebaran is much closer than the stars in the Hyades. It is 68 light-years away and is the brighter companion of a binary star pair. Located along a line about 10° east of Aldebaran is the 3.0 magnitude star ζ (zeta) Tau, marking the very tip of the bull's right horn.

Less than one degree to the north-west of zeta Tau is the well-known Crab Nebula. It is the remnant of a supernova explosion that occured in 1054 A.D. and was recorded by Chinese astronomers. Large binoculars or a small telescope shows M1 as a nebulous oval—much like the shape of a doughnut.

Another fine open cluster in Taurus is the Pleiades, or M45. Located about 8° north-west of Aldebaran, the Pleiades has six or seven stars visible to the naked eye that form a little dipper-shaped asterism. Binoculars reveal many more stars; a small telescope shows wisps of nebulous clouds surrounding as many as 100 stars. (Astronomers believe that there are 250 to 500 stars in the Pleiades.) The stars of this cluster are no more than one million years old and are still surrounded by the nebula's original gas and dust.

Near the zenith is the constellation Auriga the Charioteer. The five brightest stars of Auriga form a pentagon-shaped asterism, with Capella, or α (alpha) Aurigae, at its northernmost point. Capella, at a magnitude of 0.1 and 42 light years away, is the 6th brightest star in the sky. About 6°

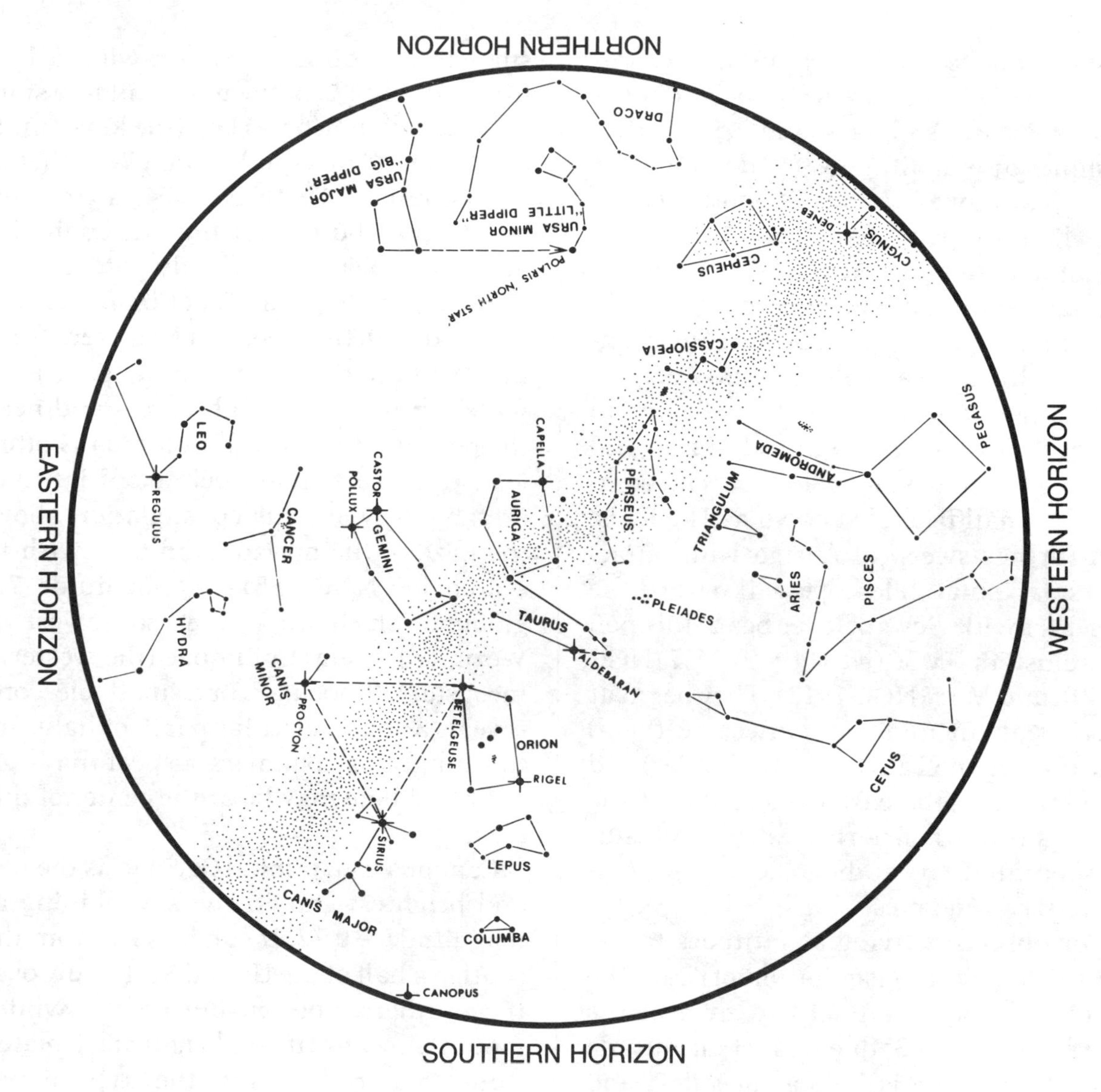

THE NIGHT SKY IN JANUARY

Latitude of chart is 34°N, but it is practical throughout the continental United States.

To use: Hold chart vertically and turn it so the direction you are facing shows at the bottom.

Chart time (Local Standard):
10 P.M. First of month
9 P.M. Middle of month
8 P.M. Last of month

Star Chart from GRIFFITH OBSERVER monthly magazine

east of Capella is β (beta) Aur, or Menkalinan—an eclipsing binary pair with a period of 3 days, 23 hours, and 2.5 minutes. It shines at magnitude 1.9 and is about 72 light-years away. About 3° south-west of Capella is ϵ (epsilon) Aurigae, an eclipsing variable with a period of 27 years—one of the longest periods known. Approximately 2½° directly south of ϵ (epsilon) Aur is another eclipsing variable, ζ (zeta) Aurigae, with a period of 2⅔ years. One degree to the east of zeta lies magnitude 3.2 η (eta) Aur. These three stars below Capella complete a small triangle known as The Kids.

A further sweep of Auriga with binoculars or a small telescope will reward the observer with views of three beautiful open star clusters—M36 (NGC 1960), M37 (NGC 2099), and M38 (NGC 1912). They begin at a point about mid-way between β (beta) Tau (just on the Taurus-Auriga border) and θ (theta) Aur (the easternmost star of the pentagon) and run northwest. Each cluster is separated by a distance of approximately two degrees.

For observers living at latitudes to the south, the small constellations of Lepus the Hare and Columba the Dove are near the meridian. About 3° due west of magnitude 3.3 μ (mu) Leporis (Lepus' brightest and northernmost star) is R Leporis, also known as "Hind's Crimson Star" after its discoverer and bright red color. R Lep is a long-period variable star that varies between magnitudes 5.5 and 11.7 over a period of 432 days. Astronomers believe this low-temperature red giant is a rare carbon star. The brightest star to the southeast is γ (gamma) Lep, a beautiful double star whose magnitude 3.7 yellow *primary*, larger and brighter, contrasts nicely with the magnitude 6.3 orange *secondary*, smaller and dimmer. By drawing a line from α (alpha) Lep (the upper-middle star) down through β (beta) Lep (the lower middle star), and an equal distance south, the observer finds M79 (NGC 1904), a 7th magnitude globular cluster that, through binoculars, appears as a cloudy patch.

The tiny constellation of Columba is associated with the dove that Noah sent from the Ark to find dry land. Because it is so far south to northern observers and near the horizon, the view of Columba is often lost. Observers living below 35° latitude are able to view this constellation above the haze of the horizon and may wish to search for NGC 1851—a magnitude 7.3 globular cluster about 8° southwest of Wezn, or β (beta) Columbae (the westernmost star). Binoculars or a small telescope reveals a small, circular patch of light. Interestingly, astronomers suspect that a giant black hole may lie at the center of this cluster.

Canopus, or α (alpha) Carinae, is the second brightest star in the sky, shining at magnitude −0.72. It can be seen from the southern half of the United States low over the southern horizon during the winter months. Because it is so bright and isolated from other bright stars in the sky, Canopus is often used as a navigational beacon to aid spacecraft going to the outer planets of our Solar System.

February

As January gives way to February, observers will begin to notice the apparent westward movement of the stars. For example, Capella has now moved west of the local meridian and will continue to move

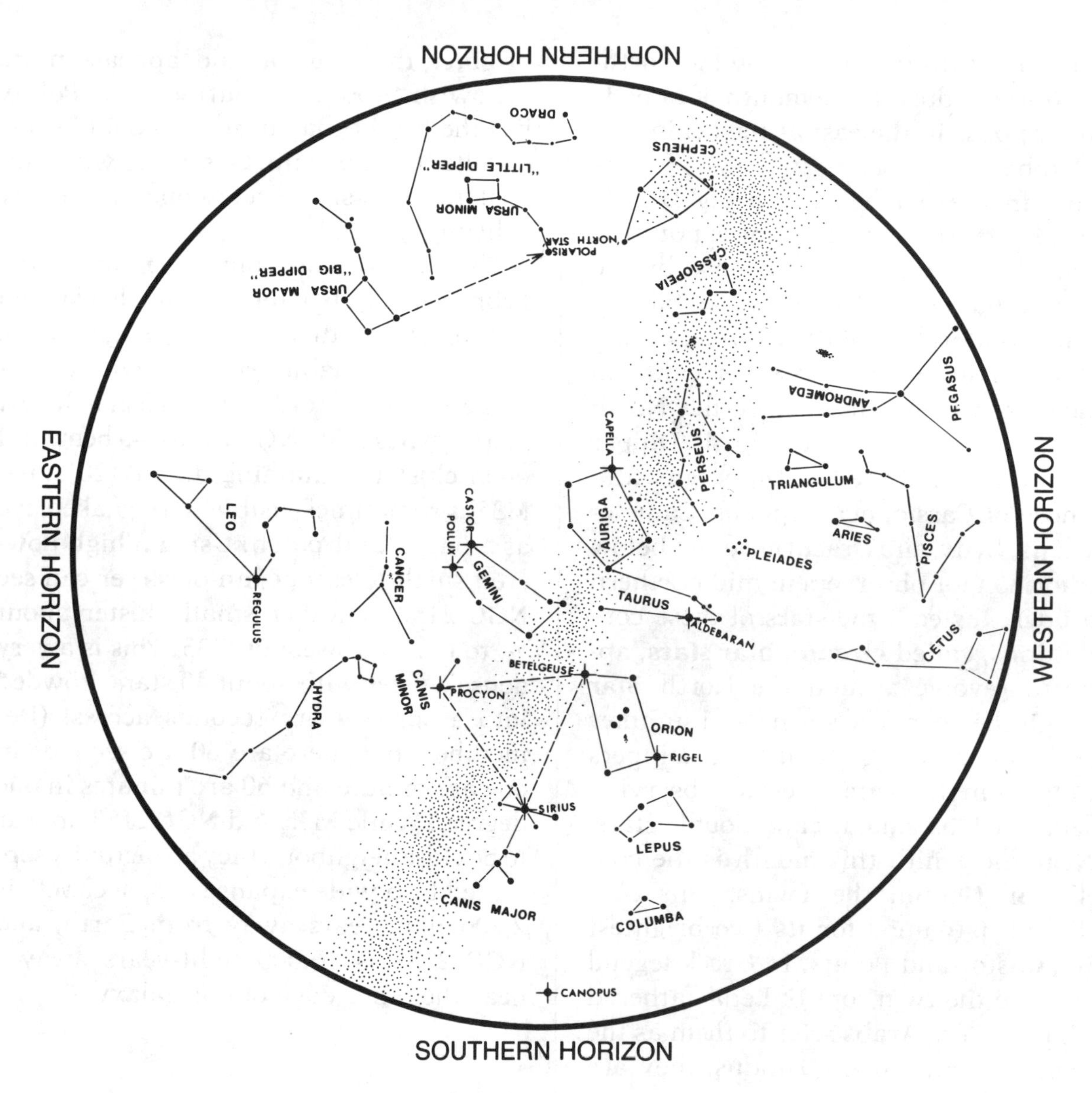

THE NIGHT SKY IN FEBRUARY

Latitude of chart is 34°N, but it is practical throughout the continental United States.

To use: Hold chart vertically and turn it so the direction you are facing shows at the bottom.

Chart time (Local Standard):
10 P.M. First of month
9 P.M. Middle of month
8 P.M. Last of month

Star Chart from GRIFFITH OBSERVER monthly magazine

west until it disappears below the northwest horizon during the month of May. It will reappear in the east at the beginning of October.

Low in the northeast, the Big Dipper hangs in the sky with its handle pointing down. By following a line formed by the two pointer stars, Dubhe, or α (alpha) UMa, and Merak, or β (beta) UMa, for 20° to the northwest, the observer can find Polaris (α Ursae Minoris), or the North Star. The stars of both the Little Dipper (Ursa Minor) and the Big Dipper (as well as those of Cassiopeia, Cepheus, Camelopardalis, Lynx, and Draco) never set below the horizon for observers in mid-northern latitudes. Instead, the stars of these constellations, called **circumpolar stars**, appear to revolve around the North Star through the year. The spring and summer months, when the Big and Little Dippers are higher in the sky, are best for observing objects such as galaxies and double stars.

Near the zenith this month is the constellation Gemini the Twins. This constellation is named for its two brightest stars, Castor and Pollux. In Greek legend they were the twin sons of Leda, fathered by Jupiter. The Arabs refer to them as the Peacocks; while to the Hindus, they are known as the Twin Deities.

The Twins is one of the few times that the fainter star Castor, or α (alpha) Geminorum, received the alpha designation, and the brighter star Pollux, or β (beta) Geminorum, received the beta designation. Castor is a beautiful double star that observers can easily resolve, or separate, with a small telescope. The two stars are magnitudes 2.0 and 2.8 and slowly revolve around each other once every 477 years. Pollux is about one-half magnitude brighter than Castor and appears more yellowish. About 5° southwest of Pollux lies the bright magnitude 3.4 double star δ (delta) Geminorum. Observers with binoculars or a small telescope will enjoy splitting this pair.

The westernmost star in Gemini on the February map is η (eta) Gem, also known as Propus. At a distance of 190 light years, η Gem is a variable red giant with an average magnitude of 3.3. About 3° northwest of η Gem lies M35 (NGC 2168)—a beautiful open cluster containing about 120 stars. M35 is sometimes visible to the naked eye as a hazy, faint patch. Using a high powered small telescope, an observer can see NGC 2158—another small cluster about ½° to the southwest of M35. This is a very dense cluster with about 40 stars crowded into a space 4 arc seconds across! (Remember that there are 60 arc seconds in one arc minute and 60 arc minutes in one degree.) While M35 and NGC 2158 appear to be close neighbors, they are actually separated by a wide expanse of space. M35 is 2,200 light-years away from Earth, and NGC 2158 is 16,000 light-years away—near the outer edge of our galaxy.

March

Although the winter stars can still be observed even as they move further west, the major spring constellations now begin to appear in the east.

Almost due south, between the constellations Gemini and Leo, lies Cancer the Crab. This constellation is not very obvious because it contains no bright stars. However, within the body of the Crab, formed by the stars γ, δ, θ, and η (gamma,

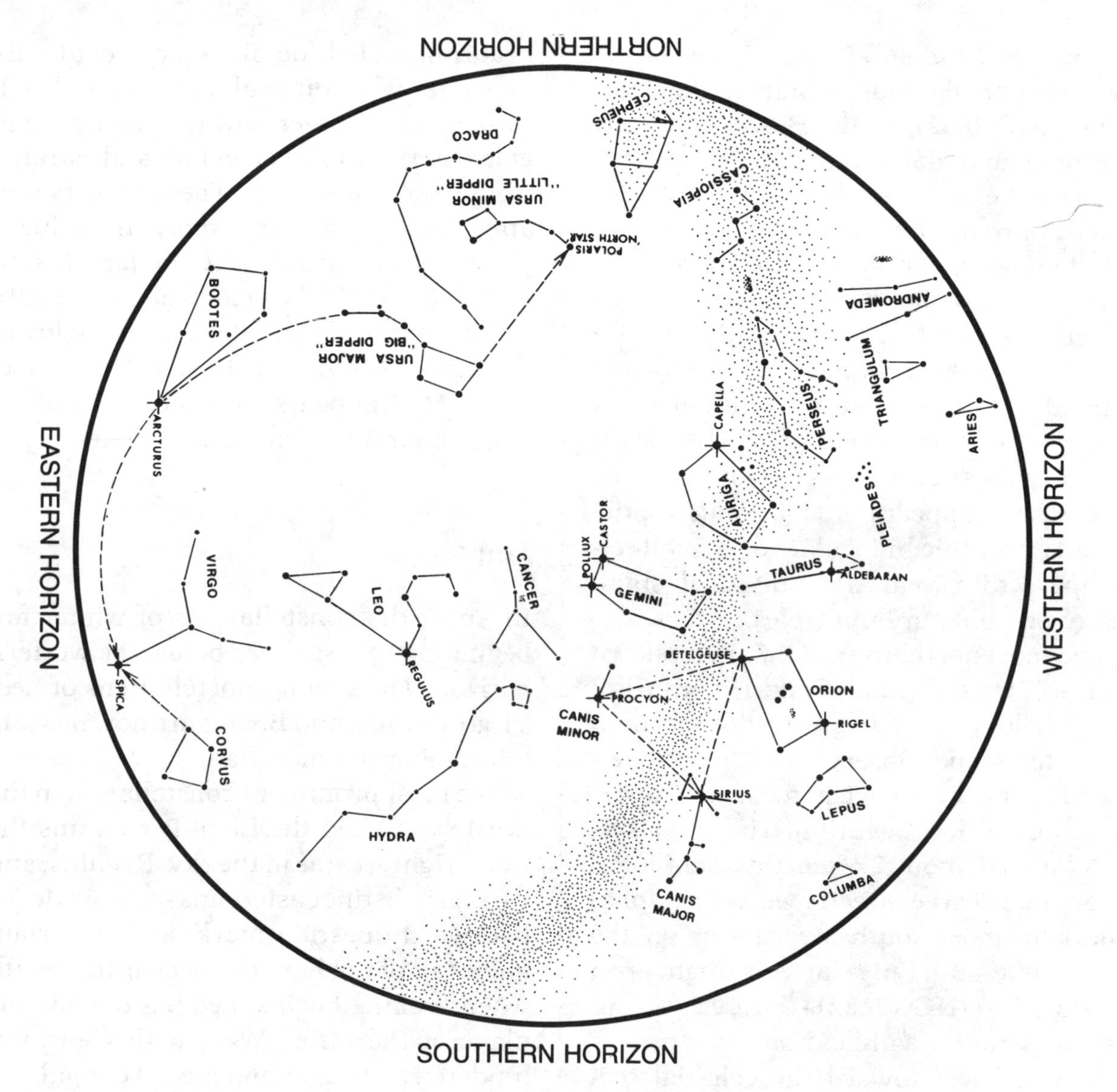

THE NIGHT SKY IN MARCH

Latitude of chart is 34°N, but it is practical throughout the continental United States.

To use: Hold chart vertically and turn it so the direction you are facing shows at the bottom.

Chart time (Local Standard):
10 P.M. First of month
9 P.M. Middle of month
8 P.M. Last of month

Star Chart from GRIFFITH OBSERVER monthly magazine

delta, theta, and eta) Cancri, is one of the most spectacular star clusters in the sky—M44 (NGC 2632), or the Beehive Cluster. (Gamma and delta are the two central stars on the chart, while theta and eta, located approximately 3° west of γ and δ Cancri, do not appear on this chart.) M44, estimated to be 520 light-years away, is an irregular patch just to the southwest of γ Cnc. (Cnc is the abbreviation for Cancri.) This cluster is sometimes referred to as Praesepe, or the Manger, with the stars γ and δ Cnc called the asses from a Greek legend. M44 appears as a hazy patch to the naked eye, while binoculars or a small telescope will reveal its individual stars, many of which are multiples.

The most northern star of Cancer shown on the chart is ι^1 (iota^1) Cancri, a beautiful and striking orange-green double in a small telescope. Observers with larger telescopes may also wish to search for ι^2 (iota^2) Cancri, a beautiful triple star system located about 2½° northwest of iota^1. About one degree directly west of α (alpha) Cnc (the most southeastern star on the chart), is a small but relatively bright open cluster, M67 (NGC 2682)—best viewed under moderate magnification.

Moving west toward the celestial meridian is the longest of the constellations—Hydra the Water Serpent. With its head near Canis Minor and Cancer, and its tail near Libra, Hydra stretches more than 100° across the southern sky. Most of the stars in this constellation are faint. The brightest is a magnitude 2.0 multiple star called Alphard, or α (alpha) Hydrae, which means "the solitary one." About 2° south of μ (mu) Hydrae (the fourth star moving west from the end of the Serpent's tail), a small telescope reveals NGC 3242—a beautiful, pale blue planetary nebula. By sweeping with binoculars or a small telescope, an observer will find many small galaxies running just below and parallel to the Serpent's body. These objects will appear to be nothing more than fuzzy glows. A magnitude 8 globular cluster, M68 (NGC 4590), is located about 4° south of 2.7 magnitude β (beta) Corvi, the lower left star of Corvus on the chart. In binoculars, M68 appears as a glowing ball of light about 3 arc minutes in diameter.

April

In April, the constellations of winter are beginning to disappear below the western horizon. The spring constellations of Leo, Virgo, Corvus, and Boötes are now nearing or crossing the meridian.

The most prominent constellation in the April sky is Leo the Lion. It contains the 21st brightest star in the sky, Regulus, and two very distinct asterisms—the sickle (or backward question mark) and the triangular tail of the lion. Leo is near the zenith when evening begins. Leo lies outside the plane of the Milky Way, with views unhindered by the gas and dust of our galaxy. Sweeping with a medium-power telescope about 9° east of Regulus, an observer can spot the outer edge of the huge Virgo Cluster of galaxies.

Regulus, or α (alpha) Leonis, sits at the base of the sickle. It is a magnitude 1.4, spectral-type B7, blue-white star at a distance of 85 light years. Visible at middle latitudes for 8 months of the year, Regulus rises a little north of east at about 9 P.M. local time on New Year's eve; it sets below the western horizon at the end of August.

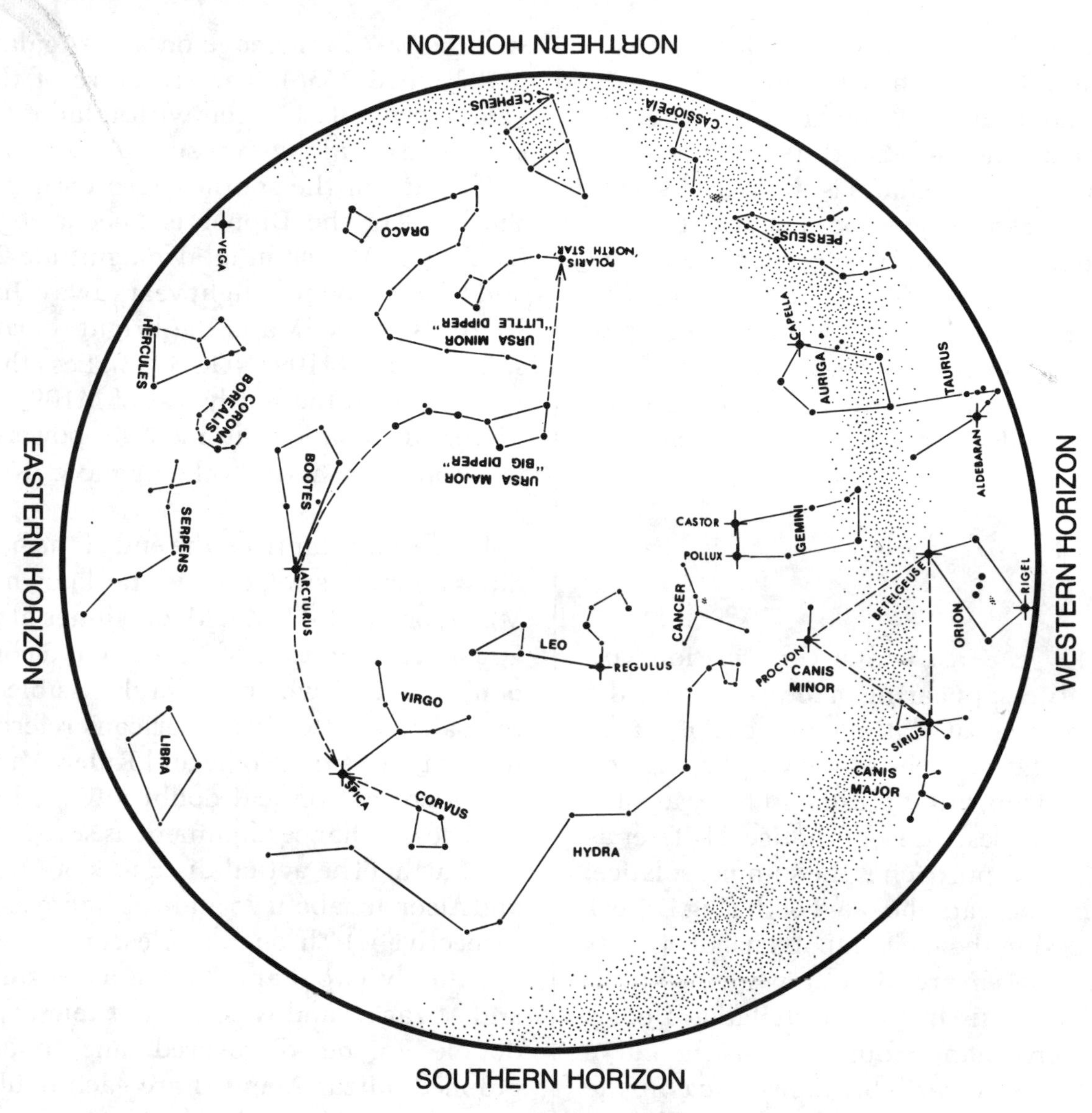

THE NIGHT SKY IN APRIL

Latitude of chart is 34°N, but it is practical throughout the continental United States.

To use: Hold chart vertically and turn it so the direction you are facing shows at the bottom.

Chart time (Local Standard):
10 P.M. First of month
9 P.M. Middle of month
8 P.M. Last of month

Star Chart from GRIFFITH OBSERVER monthly magazine

Regulus has a faint companion that can sometimes be seen with binoculars.

Located about 8° northeast of Regulus (indicated as the second star up from Regulus on the star chart) is the brightest star in the sickle, γ (gamma) Leonis. This star is a binary system composed of two yellow stars of magnitudes 2.1 and 3.4 (they are difficult to split with a small telescope). About 2° northwest of γ Leo is the radiant of the November meteor shower known as the Leonids—a shower that peaks every 33 years.

May

In May, the last winter constellations are rapidly disappearing in the west, while the summer constellations are beginning to rise in the east. The Milky Way lies around the horizon, but it is difficult to see at such low altitudes. Sunset is noticeably later as the Earth approaches the summer solstice. High overhead, the bowl of the Big Dipper is crossing the meridian. Its handle points east and then arcs down toward the bright star Arcturus in the constellation Boötes the Herdsman. From Arcturus the curve continues toward brilliant Spica in the constellation Virgo the Virgin. Finally, the curve reaches the small constellation Corvus the Crow, who rides on the back of Hydra.

Inside the four stars in the bowl of the Big Dipper lie a rich field of galaxies known as the Ursa Major cluster. About 2½° southeast of β (beta) UMa, the lower right star in the bowl, is one of the largest planetary nebulas, M97 (NGC 3587), or the Owl Nebula. (In binoculars, it is only a faint, featureless disk.) Within one degree northwest of M97 is the edge-on spiral galaxy M108 (NGC 3556). The structure of this galaxy can only be seen with a large telescope or on a long-exposure photograph.

The star in the southeastern corner of the bowl of the Dipper is Phecda, or γ (gamma) UMa, shining at magnitude 2.4 and located about 75 light years away. Just southeast of γ UMa is the bright, barred spiral galaxy M109 (NGC 3992). Less than one degree to the southwest of M109, observers using a low- to medium-powered telescope can find another galaxy, NGC 3953.

The second star from the end of the handle is an interesting double star known as Mizar, or ζ (zeta) UMa. Mizar shines at an apparent magnitude of 2.3 and has a companion, Alcor, which is barely visible to the naked eye. Native Americans referred to this pair as the Horse and Rider. Mizar and Alcor are optical doubles with their closeness a chance alignment as seen from the Earth. (The actual distances of Mizar and Alcor are about 75 and 700 light years, respectively.) Through a telescope, Mizar is actually two stars, known as Mizar A and Mizar B, and was the first telescopic double to be discovered and photographed. Mizar A and B are each double stars, but can only be detected spectroscopically—making Mizar a quadruple star system!

Crossing the meridian at this time is the constellation Virgo, associated with Astraea, the Greek goddess of justice, and with Ceres, the Roman goddess of the harvest. (The bright star Spica, or α [alpha] Virginis, stood for a stalk of grain in Ceres' hand and was often referred to as the star of prosperity.) Spica is the 16th brightest star in the sky. It is an extremely hot, bluish

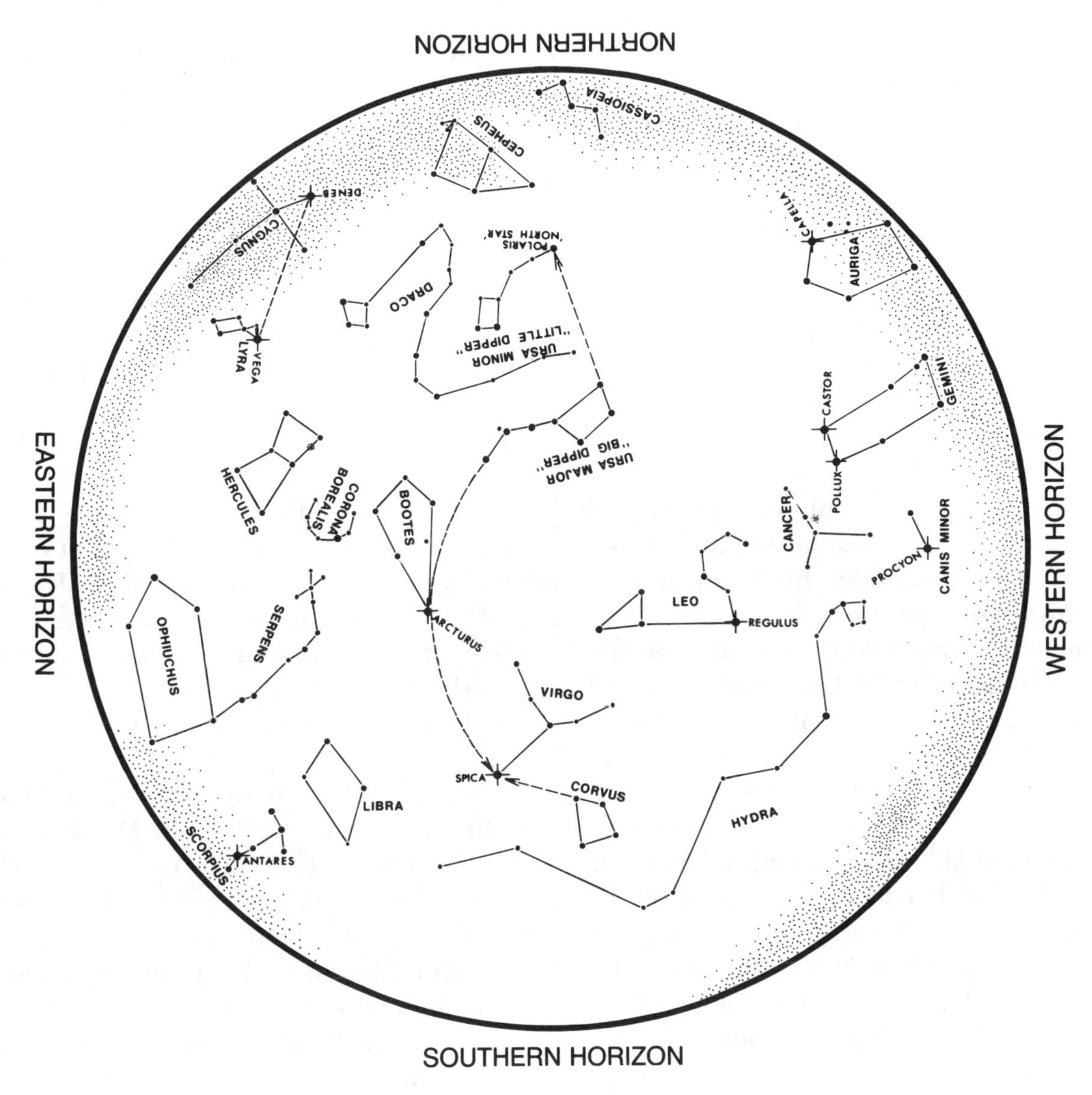

THE NIGHT SKY IN MAY

Latitude of chart is 34°N, but it is practical throughout the continental United States.

To use: Hold chart vertically and turn it so the direction you are facing shows at the bottom.

Chart time (Local Standard):
10 P.M. First of month
9 P.M. Middle of month
8 P.M. Last of month

Star Chart from GRIFFITH OBSERVER monthly magazine

star and shines at magnitude 1.0. As an eclipsing binary, Spica's intensity changes by approximately one magnitude every four days.

Five stars form a semi-circle in Virgo—from left to right, ϵ (epsilon) Vir, δ (delta) Vir, γ (gamma) Vir, η (eta) Vir, and β (beta) Vir—and contains the closest and largest cluster of galaxies known. The Virgo Cluster is only about 65 million light-years from the Earth and contains hundreds of galaxies, about 75 percent of which are spirals. In a small telescope most of these galaxies appear as small oval patches of light. However, some details of their structure are visible under high power in a medium-size telescope.

One of the more famous galaxies in the Virgo cluster is on the border of Virgo and Corvus the Crow (the small constellation south of Virgo). It is the edge-on galaxy M104 (NGC 4594), or the Sombrero Galaxy. (An edge-on galaxy is tilted to our line of sight so that we are looking sideways at the galaxy rather than down [face-on] at one of its poles, similar to viewing a plate from the side.) Located about 5° northeast of δ (delta) Corvi (the top left star in Corvus), the galaxy's dark dust lane is barely visible with a small telescope. With a long-exposure photograph, the bright galaxy's dust lane is very visible—giving the object its nickname.

Porrima, or γ (gamma) Vir (the middle star of the semi-circle), is an excellent example of a double star. Shining at a combined magnitude of 2.8, this star system is approximately 36 light-years from the Earth. Its two component stars are nearly identical spectral class F and shine at magnitudes 3.6 and 3.7. Both stars revolve around a common center of gravity every 180 years.

June

In June, the winter stars are far below the western horizon and the stars of summer are quickly appearing over the eastern horizon. The handle of the Little Dipper now points directly north; while the handle of the Big Dipper points toward the south. The constellations Boötes, marked by brilliant Arcturus, and Corona Borealis to the east are crossing the meridian high overhead. In the north, Draco's tail has already crossed the meridian; while in the south, Libra is approaching it. This is also a month of long days and short nights, so observers must wait longer for the sky to darken.

Toward the northern horizon, the bowl of the Little Dipper in Ursa Minor is on the meridian. This asterism is clearly marked by β (beta) UMi, γ (gamma) UMi, η (eta) UMi, and ζ (zeta) UMi. Stars β (beta) UMi and γ (gamma) UMi are also known as Kochab and Pherkad, respectively, and in legend, are the guardians of Ursa Minor.

The most prominent constellation is Boötes the Herdsman. Although seen as a herdsman by the ancients, most observers now say Boötes is shaped like a kite or an ice-cream cone! Arcturus, or α (alpha) Boötis, is the 3rd brightest star in the sky at magnitude 0.0, and is approximately 36 light years from the Earth. Its spectral type is K2, and it has a diameter about 25 times that of the Sun. Arcturus is also one of the fastest moving stars with a radial velocity of more than 80 miles per second. To ap-

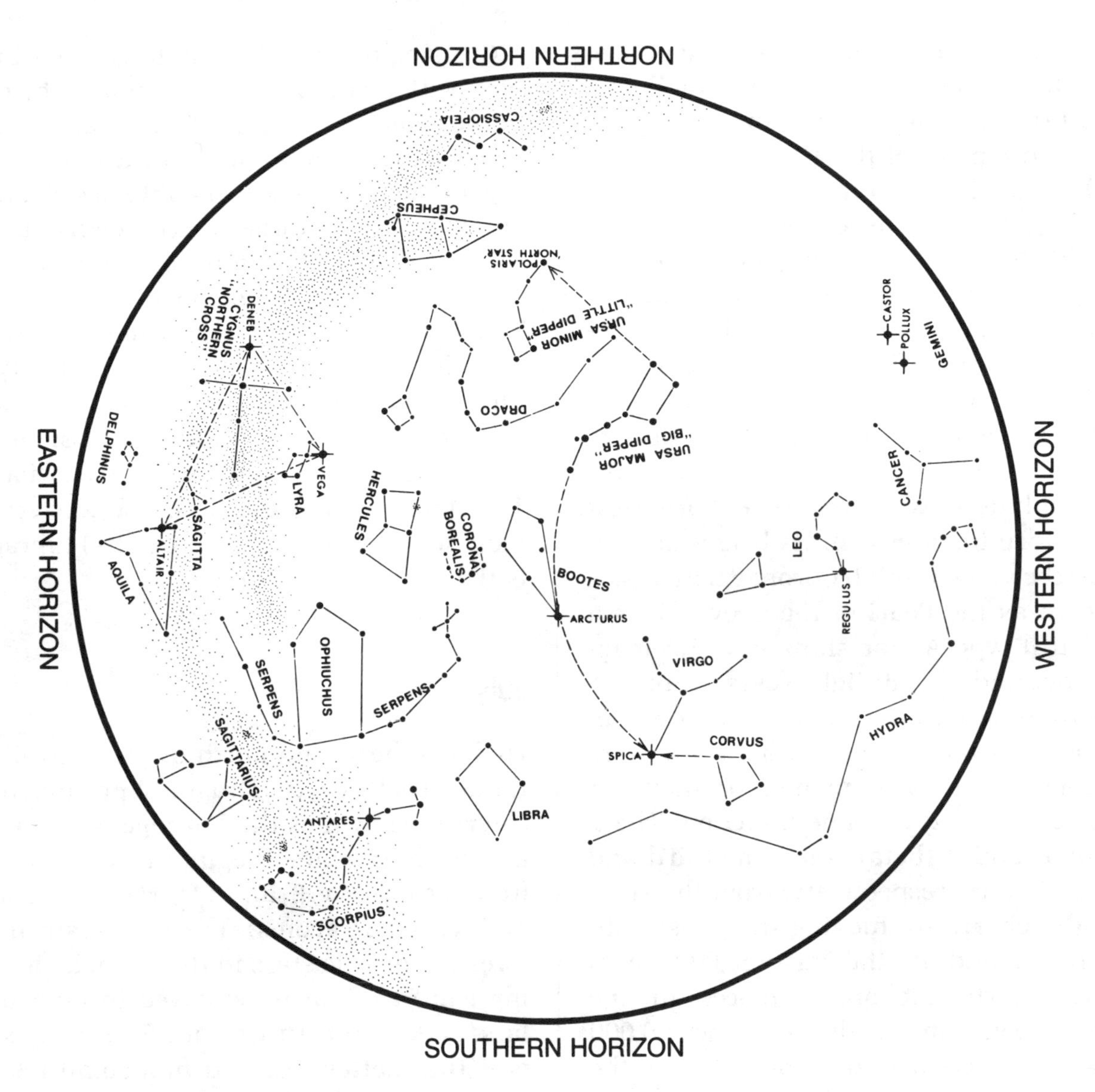

THE NIGHT SKY IN JUNE

Latitude of chart is 34°N, but it is practical throughout the continental United States.

To use: Hold chart vertically and turn it so the direction you are facing shows at the bottom.

Chart time (Local Standard):
10 P.M. First of month
9 P.M. Middle of month
8 P.M. Last of month

Star Chart from GRIFFITH OBSERVER monthly magazine

preciate the great speeds and distances of the stars, in 800 years as seen from Earth, Arcturus will only move a distance equal to the diameter of the Moon.

The star above Arcturus on the left side of the kite is ϵ (epsilon) Boo, a beautiful double whose two members are orange and green. About 4° northeast of δ (delta) Boo (the next star above epsilon on the chart), is μ (mu) Boo—a white and orange pair with even fainter companions.

Corona Borealis, or the Northern Crown, lies just east of Boötes. It is a distinctive constellation, with the seven brightest stars forming a crown. Its brightest star, Alphecca, or α (alpha) Coronae Borealis, is known as the Pearl of the Crown. It is a spectral-type A star shining at 2.2 magnitude and is 78 light-years from the Earth. Again, this constellation is only an optical pattern; none of the seven stars are gravitationally related or move at the same speed or same direction. The α (alpha) and β (beta) stars (the third and second stars, respectively, from the right on the chart) are moving in opposite directions, and in the past 50,000 years have nearly exchanged places in the sky as seen from Earth. In another 50,000 years, as the stars continue to move, the constellation will no longer resemble a crown.

An interesting variable star is 2° north of δ (delta) CrB (the third star down from the left on the chart). Normally this star, R Coronae Borealis, shines at 6th magnitude for months and even years, but then it quickly drops to 11th magnitude (or fainter) for only a few weeks. Astronomers generally agree that R CrB is a carbon-rich star, periodically throwing off clouds of carbon soot and dimming its photosphere.

Crossing the meridian in the south is Libra the Balance, the only inanimate object in the zodiac. Its two brightest stars—Zubenelgenubi, or α (alpha) Librae (a widely separated double star), and Zubeneschamali, or β (beta) Librae—once belonged to the constellation Scorpius to the southeast and were known as Chelae, or the Claws. Their names mean the "Southern Claw" and the "Northern Claw," respectively. The Romans created Libra to mark the time of the autumnal equinox and assigned these two stars to the new constellation. The star α (alpha) Librae is the westernmost star on the chart and β (beta) Librae is the northernmost star.

July

High overhead and north of the zenith is the head of Draco the Dragon. A prominent asterism known as the Lozenge is formed by the four stars (beginning clockwise from the top) ξ, ν, β, and γ Dra (xi, nu, beta, and gamma Draconis). There are many interesting double stars in this area, including ν(nu) Dra, easily resolved in binoculars, and μ (mu) Dra, about 5° northwest of β Dra, better resolved in a small telescope. Located where the body of the dragon turns sharply to the west (the second star up from the Lozenge), is ϵ (epsilon) Dra—a good double star separated by only 3 arc seconds and resolved with a telescope at medium magnification.

Midway on a line connecting δ (delta) Dra (the first star up from the Lozenge) and ζ (zeta) Dra (the sixth star to the west on the dragon's body), lies one of the brightest planetary nebulae—NGC 6543. At magnitude 8.8, a small telescope reveals

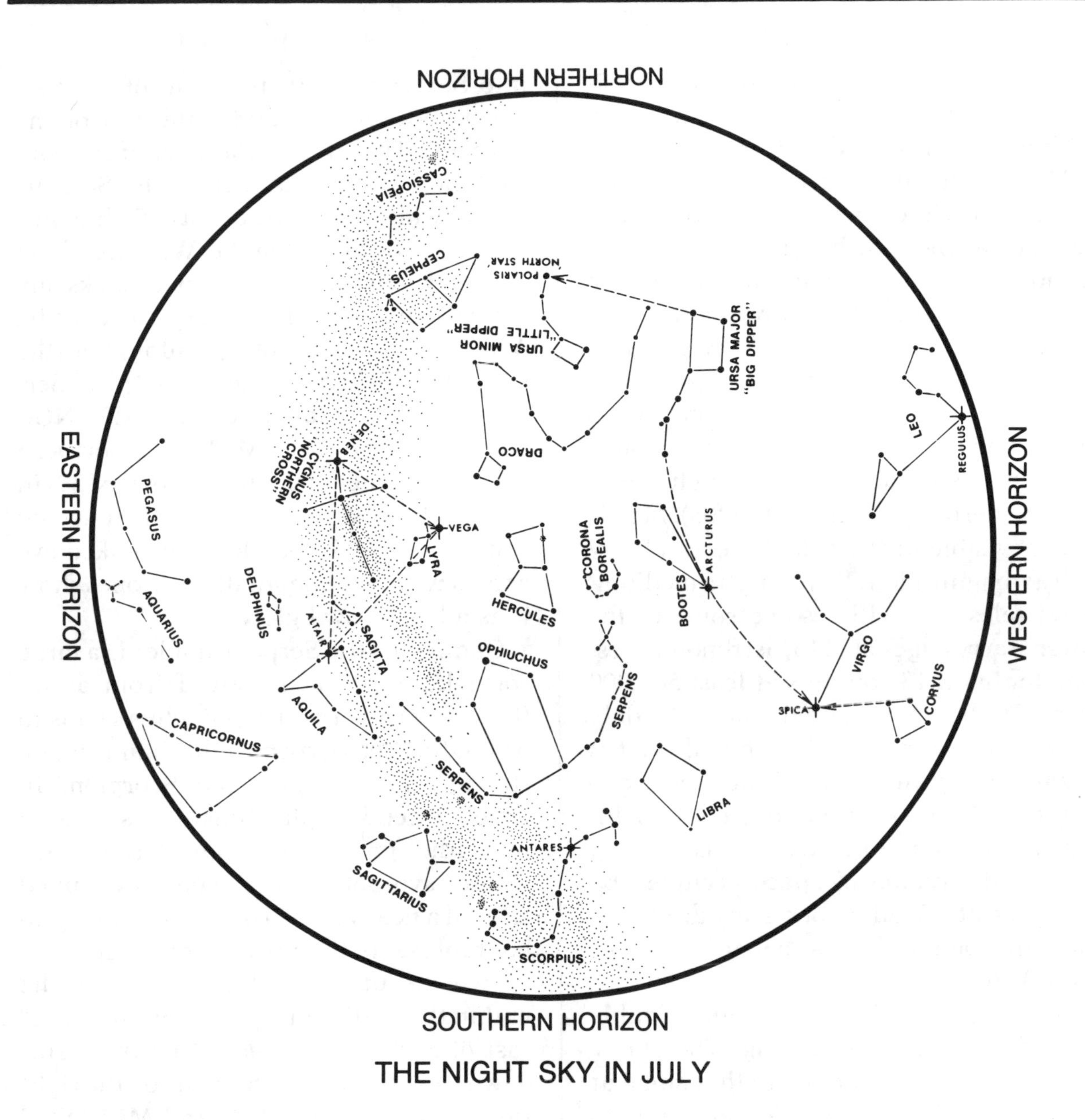

THE NIGHT SKY IN JULY

Latitude of chart is 34°N, but it is practical throughout the continental United States.

To use: Hold chart vertically and turn it so the direction you are facing shows at the bottom.

Chart time (Local Standard):
10 P.M. First of month
9 P.M. Middle of month
8 P.M. Last of month

Star Chart from GRIFFITH OBSERVER monthly magazine

a fuzzy disk. A more powerful telescope is needed to reveal any internal structure, such as the magnitude 9.6 central star.

The constellation Hercules, named for the famous Greek hero, is now high overhead, or at the zenith. The four stars η, ζ, ϵ, and π (eta, zeta, epsilon, and pi) Her form a prominent asterism known as the Keystone. About a third of the way south on a line connecting η (eta) and ζ (zeta) Her (the two stars, respectively, from top to bottom on the right side of the Keystone), is the most spectacular globular cluster in the northern sky, M13 (NGC 6205). M13 is faintly visible to the naked eye as a fuzzy ball at magnitude 4.0. Although a medium-sized telescope will resolve some of the outer starry edges of M13, astronomers estimate that M13 contains at least 500,000 stars. It should be noted that M13 only appears to be in Hercules. In reality, it is 25,000 light-years away—much more distant than the stars of the constellation. Interestingly, our Solar System is rushing toward the region of space occupied by Hercules at 12 miles per second. M13 is indicated on the chart as a star just below η (eta) Her.

Hercules contains many fine double stars for amateur observing. The star ζ (zeta) Her has two members that shine at magnitudes 2.9 and 5.5 and are only 30 light-years away from the Earth. (On the chart ζ (zeta) Her is the star directly below M13.) Another fine double is δ (delta) Her, the most southeastern star on the chart. Many different colors have been used to describe this double star.

South of Hercules is the constellation of Serpens the Serpent, which stretches nearly 60° across the southern sky. Astronomers have divided this constellation into Serpens Caput (the head of the Serpent) and Sepens Cauda (the tail of the Serpent). The stars to the west of the constellation Ophiuchus belong to Serpens Caput; while those to the east of Ophiuchus belong to Serpens Cauda. With its alpha star only magnitude 2.7, Sepens lacks any bright stars to make it easy to identify. About 7° to the west of ϵ (epsilon) Ser (the sixth star down, near the letter "p" of Serpens), is the globular cluster M5 (NGC 5904), rivaled only by M13 in the northern sky and ω (omega) Centauri (NGC 5139) in the southern sky. Under excellent seeing conditions, M5 is visible to the naked eye as a hazy star, while small telescopes show it as a bright oval glow.

Ophiuchus the Serpent-holder is a large constellation. When viewed from above the southern horizon, Ophiuchus seems to be holding the Serpent in his hands while standing on Scorpius the Scorpion. In Greek legend, Ophiuchus is associated with the great physician Aesculapius, whose symbol was a snake wrapped around a heavy staff. To this day, this symbol represents the medical profession.

Two beautiful globular clusters lie within the outline of Ophiuchus. About 10° east of δ (delta) Ophiuchi (the third star down from the star at the peak, on the right side) is M10 (NGC 6254) and M12 (NGC 6218). These two globular clusters are very similar and are visible to the naked eye under the best viewing conditions. M12 lies about 3° northwest of M10 and is a little larger. A third globular cluster, M107 (NGC 6171), is located about 3° south of the mid-way point of a line drawn between δ (delta) Ophiuchi and the star to its east, η (eta) Ophiuchi. It is smaller and fainter (magnitude 9) than M10 or M12.

Another fine globular cluster, M19 (NGC 6273) at magnitude 6.9, lies 8° east of Antares (see below). To find M19, sweep the telescope east and slightly north, or point it at Antares and wait 33 minutes for the Earth's rotation to bring M19 into view. Another small cluster, M62 (NGC 6266), is about 5° south of M19. In a telescope under high power, the nucleus of M62 looks like an elongated comet.

Crossing the meridian low in the southern sky is the constellation Scorpius the Scorpion. Its S-shaped body is very prominent in the direction of the southern Milky Way's central disk. Observations of this region of the summer sky reveal a field rich in star clusters, bright nebulae, and dark, winding lanes of dust. A good practice is to sweep the area with binoculars or a telescope at low power, then change to higher magnification for individual objects.

Antares, or α (alpha) Scorpii, is the bright red eye of the Scorpion. It is the 15th brightest star, and its color is often compared to the red planet Mars, thus earning it its name Antares or "compared with Mars." At magnitude 1.0, Antares is easily visible. It is a variable red giant of spectral type M (cool, red stars) and is located about 330 light years away.

About 2° due west of Antares is M4 (NGC 6121), one of the largest and nearest globular clusters. M4 shines at magnitude 6.0 and is sometimes visible to the naked eye; when viewing M4 through a telescope, be sure to remove Antares' bright light from the telescope's field of view. For observers using higher powers, another smaller globular cluster, NGC 6144, lies just under 1° northeast of M4. Mid-way on a line drawn between Antares and β (beta) Sco (the westernmost star in the middle of the claw), is another small but bright globular cluster, M80 (NGC 6039).

Two large, bright open star clusters lie northeast of the Scorpion's stinger, marked by the stars λ (lambda) Sco, or Shaula, at magnitude 1.6, and υ (upsilon) Sco, or Lesath, at magnitude 2.7. (These are the two stars on the end of the Scorpion's body at the left.) The first open cluster, M7 (NGC 6475), is very bright at magnitude 5.0 and contains many 6th to 8th magnitude stars. It is one of the few open clusters that look spectacular in binoculars. On the northwest edge of M7 is the globular cluster NGC 6453, best seen under moderate to high power. About 3° northwest of M7 is another globular cluster, M6 (NGC 6405), that shines at a total magnitude greater than 6.0. It contains many stars between magnitudes 8.0 and 12.0. (M6 and M7 appear on the chart as oval patches northeast of the stinger stars.)

Around ζ (zeta) Sco (the fourth star down from Antares) observers far enough south to see this region will find several more deep sky objects. An open cluster, NGC 6231, is about ½° north of ζ (zeta) Sco and looks like a miniature version of the Pleiades. It is about 5,700 light-years away. About 1½° north of ζ (zeta) Sco lies a rich cluster of stars known as H12, where 200 stars extend northeast from NGC 6231 and end in a bright nebulous patch.

August

This month the summer Milky Way is high overhead, bringing with it two constellations rich with stars—Lyra the Lyre, and Sagittarius the Archer. The other constellations of summer—Cygnus the Swan, Sagitta the Arrow, Aquila the Eagle, and Delphinus the Dolphin—are rapidly approaching the meridian and will be discussed in September's observations.

Lyra is viewed on the chart as a small parallelogram at the zenith, with its brightest star, α (alpha) Lyrae, or Vega, on a line extending to the northwest. Vega shines brilliantly at magnitude 0.0 and is the 4th brightest star in the sky. (In 1850, Vega became the first star to be photographed.) Vega was once the Earth's north (or pole) star; in 12,000 years, it will be in that position again due to the precession of the Earth's axis (see Chapter 2). Vega also forms a small equilateral triangle with the stars ε (epsilon) and ζ (zeta) Lyrae. An observer with very good eyesight might be able to detect the two components of ε (epsilon) Lyr, each at magnitude 5.0. A small telescope easily separates this pair and splits each star into another pair—thus making ε (epsilon) Lyr a quadruple star system.

The bottom right star in the parellelogram is β (beta) Lyr, a spectroscopic double and eclipsing binary whose magnitude varies from 3.4 to 4.3 every 12.9 days. It is interesting to follow the differences in the brightness of this star by comparing it to nearby magnitude 3.2, γ (gamma) Lyr less than 2½° to the southeast. Located about halfway between β and γ Lyr is M57 (NGC 6720), or the Ring Nebula. This nebula is one of the most famous planetary nebulae, named because their shapes resemble a planet. In reality, the ring is the ejected shell of a star's outer mass that exposes the star's core. (This will be the fate of our Sun in about another 5 billion years.) A small telescope reveals the doughnut-shaped ring clearly, although a larger telescope is necessary to show any structural detail. Only very large instruments give a glimpse of the 15th magnitude, dying central star.

Low in the southern sky is Sagittarius the Archer. Although a figure from classical Greek mythology, to a modern observer, Sagittarius resembles a teapot. Its three brightest western stars are usually referred to as the spout of the teapot and point in the direction toward the center of our galaxy. This constellation is so rich in deep-sky objects that an entire book could be written about them alone! The northernmost star, λ (lambda) Sagittarii, or Kaus Borealis, at magnitude 2.8, is a very good reference point for exploring this region of space.

About 5° northwest of λ (lambda) Sag, is the bright emission nebula M8 (NGC 6523/30), or the Lagoon Nebula. Visible to the naked eye as a comet-like glow, this cloud of gas and dust contains hundreds of young and newly forming stars that cause the nebula to light up. Because M8 appears as large as the full Moon, it is best to observe it with binoculars or a telescope with a wide-field eyepiece. Just 1½° to the north-northwest of M8 is M20 (NGC 6514), or the Trifid Nebula. Visible in a moderate telescope at medium power, the three areas separated by dark lanes of dust give this emission nebula its name. M8 and M20 are indicated on the chart as the two oval patches northwest of λ (lambda) Sag. About 10° north of λ (lambda) Sag is M17 (NGC 6618), also known as the Omega Nebula or Swan Nebula because of its graceful curved shape.

The rich open cluster M23 (NGC 6494) lies about 4° north-northwest of M20. Discovered by Messier in 1764, this cluster

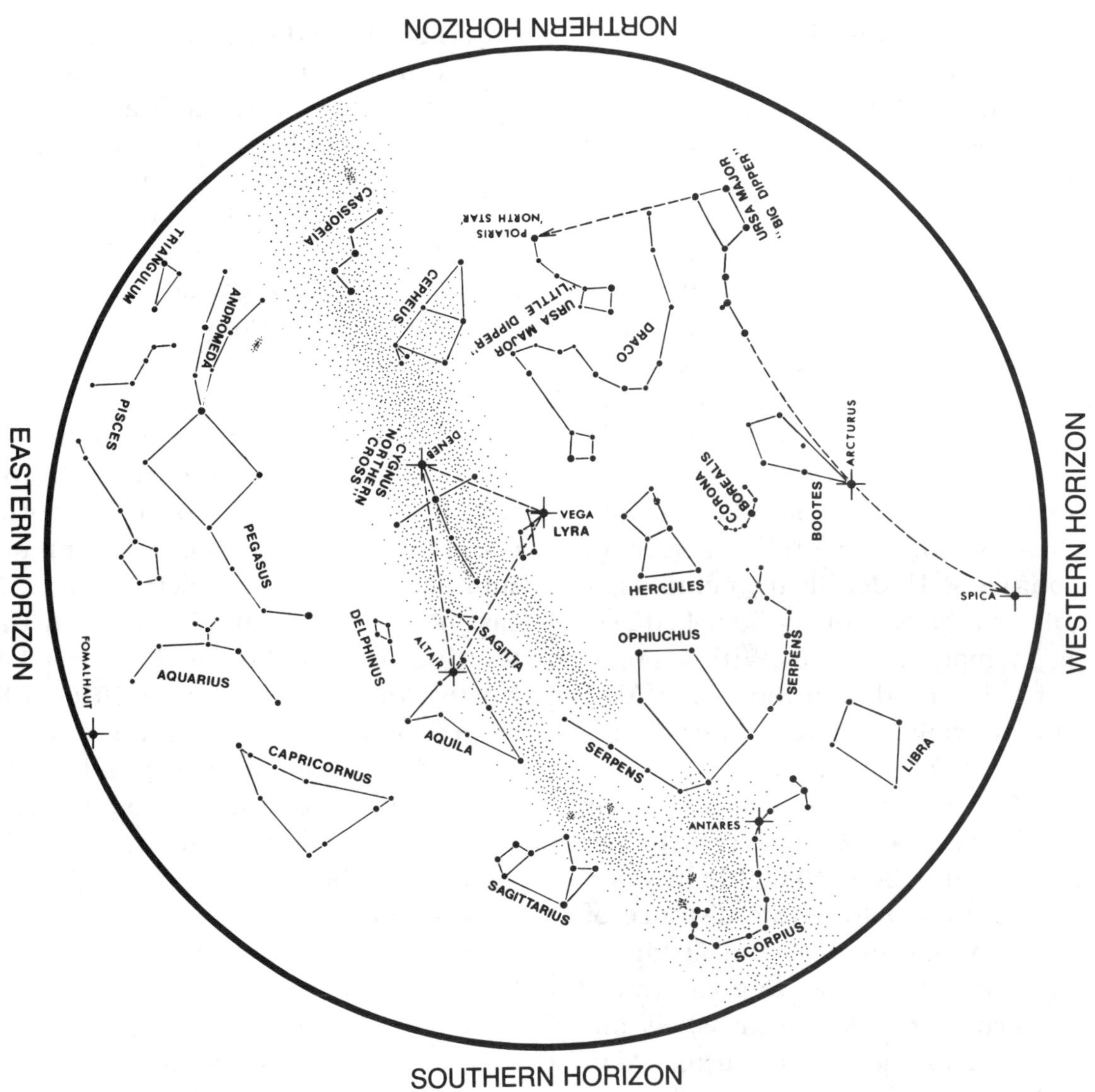

THE NIGHT SKY IN AUGUST

Latitude of chart is 34°N, but it is practical throughout the continental United States.

To use: Hold chart vertically and turn it so the direction you are facing shows at the bottom.

Chart time (Local Standard):
10 P.M. First of month
9 P.M. Middle of month
8 P.M. Last of month

Star Chart from GRIFFITH OBSERVER monthly magazine

contains about 120 stars in an area the size of the full Moon. M21 (NGC 6531) is another small, rich open cluster at magnitude 6.5, about ½° northeast of M20. It can be seen in the same field of view as M20. A compact globular cluster, M75 (NGC 6864), lies about 15° northeast of α (alpha) Sag (the top star of the handle) near the border of Capricornus. It is one of the most distant Messier objects at 95,000 light-years from the Earth. This cluster shines at 8th magnitude and is easily visible in good binoculars.

Another impressive globular cluster, M22 (NGC 6656), is about 2° northeast of λ (lambda) Sag. Under ideal conditions, it is visible as a hazy spot to the naked eye, shining at magnitude 5.0. With a population of at least half a million stars, M22 is the third brightest and most easily resolved globular cluster in the sky—rivaled only by M13 in Hercules. The reason for such visibility is M22's nearness to the Earth. It is only about 10,000 light-years away as compared to M13's distance of 24,000 light-years. In eastern Sagittarius, about 7° east of ζ (zeta) Sag (the most southeastern star in Sagittarius), is another spectacular globular cluster, M55 (NGC 6809). It is easily seen by the naked eye and looks like a hazy star in binoculars. In a small telescope, M55 looks like a hazy circle about 10 arc minutes across. Unfortunately, it is somewhat difficult to resolve for observers in northern latitudes because it is close to the horizon.

September

As September begins and dusk is noticeably earlier, the constellations of autumn are rising above the eastern horizon. Crossing the meridian are the last constellations of summer: Cygnus, Sagitta, Delphinus, and Aquila. Capricornus the Sea-Goat, low in the southern sky, rounds out this month's areas of stellar interest.

High overhead at the zenith is Cygnus the Swan, with magnitude 1.3 Deneb, or α (alpha) Cyg, marking its tail. Most observers find Cygnus' other identity as the Northern Cross easier to see against the Milky Way's rich star fields. A line connecting Deneb west to Vega, south to Altair, and then back up to Deneb, outlines one of the most prominent asterisms of the summer sky—the Summer Triangle. Just east of Deneb is a large area of nebulae that includes the famous (but very faint) North American Nebula (NGC 7000), named because its shape resembles the North American continent; and the Pelican Nebula (IC 5067–5070). Observers with wide-field binoculars may be able to distinguish these nebulae from the starry background. Another area rich in diffuse nebulae is around Sadr, or γ (gamma) Cyg—the star in the center of the cross. It is best to view this area with binoculars or a telescope at low power.

To the southeast of Sadr and marking the end of the left arm of the cross, is magnitude 2.5 ε (epsilon) Cyg. The beautiful Veil Nebula (NGC 6960, 6992–6995, and 6979) is just 2½° south of this star. The remnant of a supernova explosion over 10,000 years ago, the Veil Nebula is visible in both binoculars and telescopes with wide-field, low-power eyepieces.

Marking the nose of the Swan in the southwest of the constellation is one of the most beautiful double stars—Albireo, or β (beta) Cyg. Binoculars split this pair into

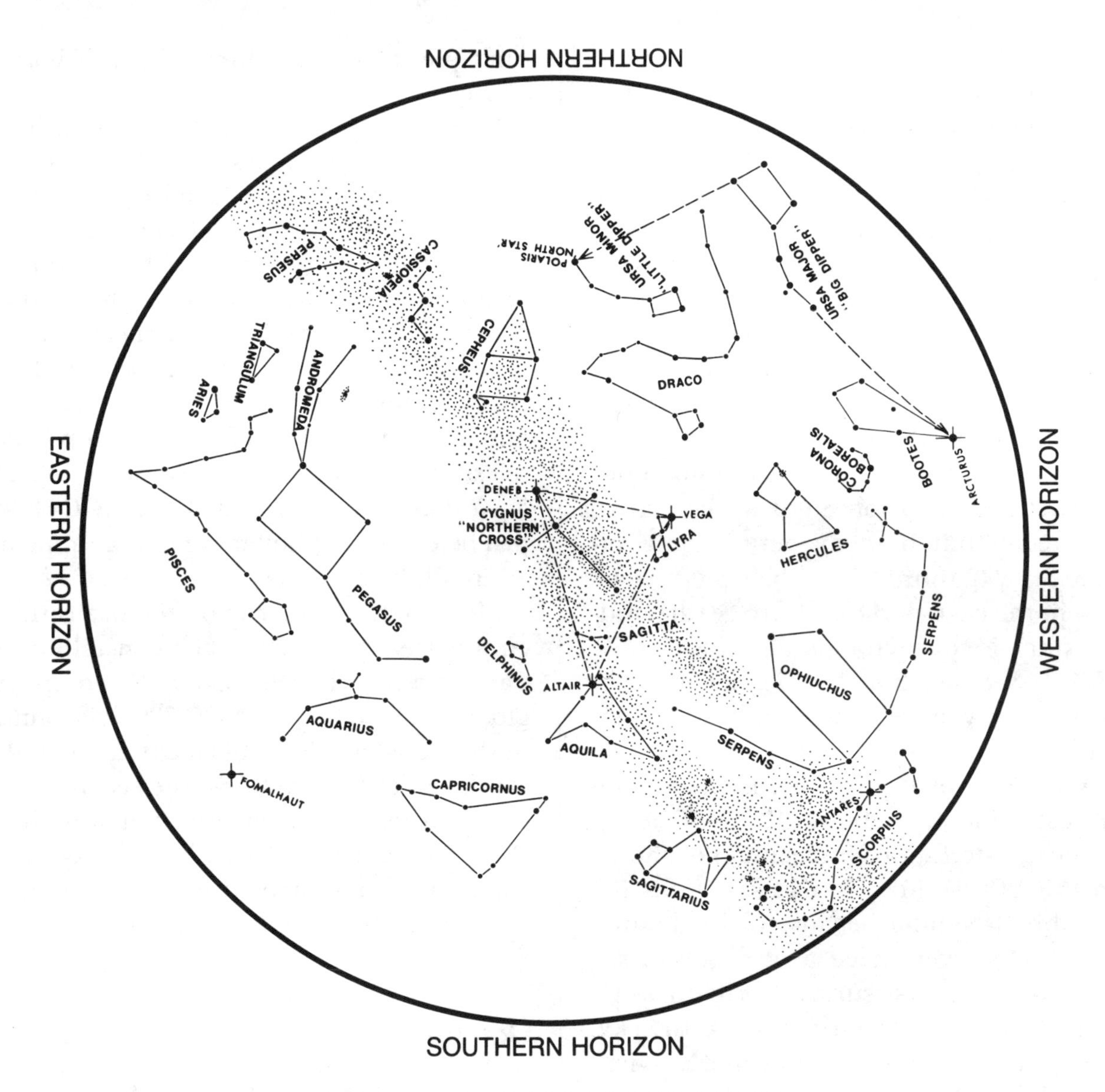

THE NIGHT SKY IN SEPTEMBER

Latitude of chart is 34°N, but it is practical throughout the continental United States.

To use: Hold chart vertically and turn it so the direction you are facing shows at the bottom.

Chart time (Local Standard):
10 P.M. First of month
9 P.M. Middle of month
8 P.M. Last of month

Star Chart from GRIFFITH OBSERVER monthly magazine

its orange and blue stars that shine with a combined magnitude of 3.1. Albireo is a good subject for viewing color contrast between stars.

The Milky Way appears to divide into two separate lanes south of Cygnus into Aquila. In reality, clouds of dust form a long band along the central plane of our galaxy, known as the Great Rift, that lies along the central plane of our galaxy (like the dust lane seen in the edge-on Sombrero Galaxy, M104). (See observations for May.) Just east of the Great Rift is the small but distinct constellation Sagitta, the Arrow. Located about halfway between magnitude 3.5 γ (gamma) Sge (the tip of the arrow in the east) and δ (delta) Sge (the middle star), is the rich globular cluster M71 (NGC 6838). Although M71 is relatively small, many stars can be seen using binoculars or a small telescope.

About 3° north of γ Sge, is the second largest planetary nebula, M27 (NGC 6853), in the constellation Vulpecula (not shown on this chart). M27 is also known as the Dumbbell Nebula because of its distinct shape. Observers need a dark sky to see M27 because its surface brightness is spread over such a large area of the sky. Low power reveals its shape and grayish-green color.

The small constellation Delphinus the Dolphin resembles a kite with a tail. It shouldn't be overlooked by observers as it contains several multiple stars that make it interesting for viewing. Three of the four stars in the kite are multiple stars: γ (gamma) Del, the northernmost star, is a beautiful double easily split in a small telescope; α (alpha) and β (beta) Del (moving clockwise from gamma) are the other two multiple stars in Delphinus that are worth observing.

Due south of Sagitta is the constellation Aquila the Eagle, marked by its brilliant alpha star, Altair—the 12th brightest star in the sky at magnitude 0.8. On a line from magnitude 2.7 γ (gamma) Aql (the star 2° northwest of Altair) down through δ (delta) Aql to λ (lambda) Aql (the most southwestern star) are many planetary nebulae. By sweeping along this line, observers can find almost 10 nebulae. Medium-power telescopes show two good examples, about 5° west of Altair: NGC 6803 and NGC 6804 just below. Both planetary nebulae appear as small, fuzzy disks.

Close to the southern horizon in a rather empty area of the sky is the constellation Capricornus, the Sea-Goat. A compact globular cluster, M30 (NGC 7099), is found in this constellation. It is located about 4° southeast of ζ (zeta) Cap—the second star down from the far eastern star δ (delta) Cap. This cluster looks like a fuzzy star since only large-aperture telescopes are able to resolve its individual stars.

October

During the month of October the sky is in transition. The constellations of winter are rising in the east; while in the west, the stars of summer are rapidly retreating. Crossing or near the meridian are the constellations Cepheus, Pegasus, Aquarius, and tiny Grus.

Cepheus, named for a mythological king of Ethiopia, appears high overhead at the zenith and is shaped like a triangle on top of a square. Just to the east of its south-

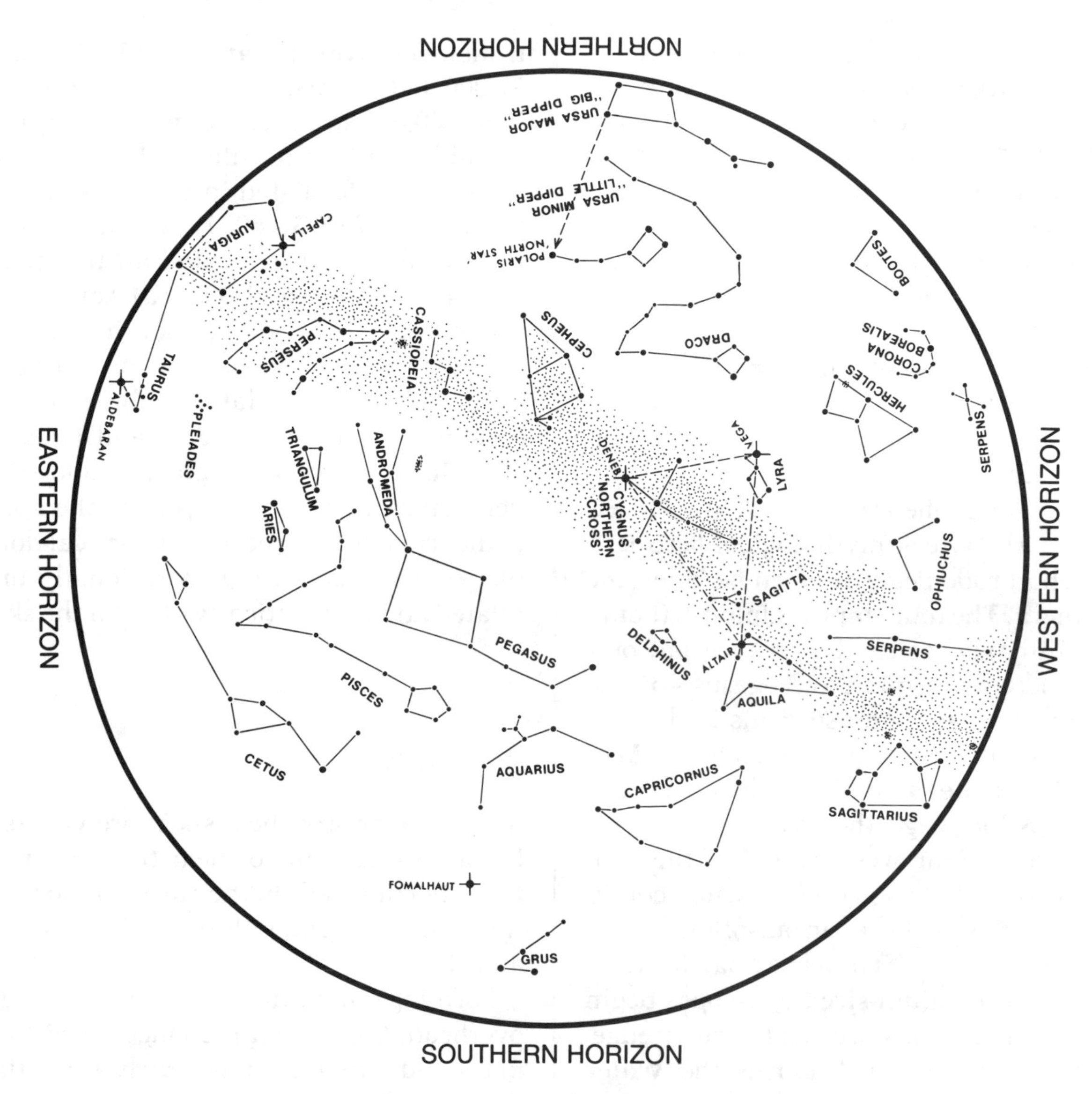

THE NIGHT SKY IN OCTOBER

Latitude of chart is 34°N, but it is practical throughout the continental United States.

To use: Hold chart vertically and turn it so the direction you are facing shows at the bottom.

Chart time (Local Standard):
10 P.M. First of month
9 P.M. Middle of month
8 P.M. Last of month

Star Chart from GRIFFITH OBSERVER monthly magazine

ernmost star, ζ (zeta) Cep, is δ (delta) Cep—a double whose brighter component is the historical prototype of the Cepheid Variables. In 5.3 days, δ Cep ranges from magnitude 3.5 to 4.4. The fainter companion shines at magnitude 7.5 and its blue contrasts nicely with the yellow of its partner. Alderamin, or α (alpha) Cep, is Cepheus' most western star at magnitude 2.4. About 52 light-years away, this star will be close to the north celestial pole in 5,000 years, replacing Polaris because of the precession of the Earth's axis.

South of Cepheus is Pegasus the Winged Horse. The Greek mythological hero Bellerophon rode Pegasus when he killed the Chimera. The four stars α (alpha), β (beta), γ (gamma) Pegasi, and α Andromedae outline the Great Square of Pegasus—one of the most prominent asterisms in the sky. The 6.4 magnitude globular cluster M15 (NGC 7078) lies about 5° northwest of Enif, or ε (epsilon) Peg—the nose of Pegasus located at the far western end of the constellation. Under prime viewing conditions, M15 is visible as an out-of-focus star. Binoculars reveal it as a hazy patch; while small and medium-sized telescopes begin to resolve the stars around the outer edge.

The area around Aquarius the Water Bearer has been known as the watery part of the sky since the time of the Babylonians. The constellation is usually depicted as a man pouring water from a jar, with the stream flowing south toward the bright star Fomalhaut (the bright star in the southern constellation of Piscis Austrinus, the southern fish). M2 (NGC 7089) is a bright globular cluster at magnitude 7.0 about 9° west of α (alpha) Aquarii (the second star from the left) and 6° north of β (beta) Aqr (the end star on the left). Astronomers have estimated that M2 is about 37,000 light-years away and contains at least 100,000 stars. This cluster is very observable with binoculars. The faintest globular cluster listed in the Messier catalogue is M72 (NGC 6981), which glows at magnitude 9.0. It is located a little more than 10° southwest of β (beta) Aqr.

For observers far enough south, Grus the Crane, appears to hover over the southern horizon. This constellation's three brightest stars—α (alpha), β (beta), and δ (delta) Gru—form an almost right triangle. The delta star (the first star up from the apex of the triangle) is a red-yellow optical double; that is, they are gravitationally unrelated and only optically close in the sky.

November

As winter approaches, six more constellations are crossing or near the meridian. From north to south, they are: Cassiopeia, Andromeda, Triangulum, Aries, Pisces, and Cetus.

Forming an unmistakeable "W" high overhead, Cassiopeia the Queen of Ethiopia, stands out almost as clearly as the Big Dipper. Because it is found in the Milky Way, Cassiopeia is rich in deep-sky objects. However, an observer may first want to take a look at γ (gamma) Cassiopei—the central star of the "W." This star is the prototype of an irregular variable star class. These variables are thought to be rapidly rotating B-class giants that randomly vary their brightness by as much as 1½ magnitudes. For example, γ (gamma) Cas fluctuates between magnitudes 1.6 and 3.0. (A high-powered tele-

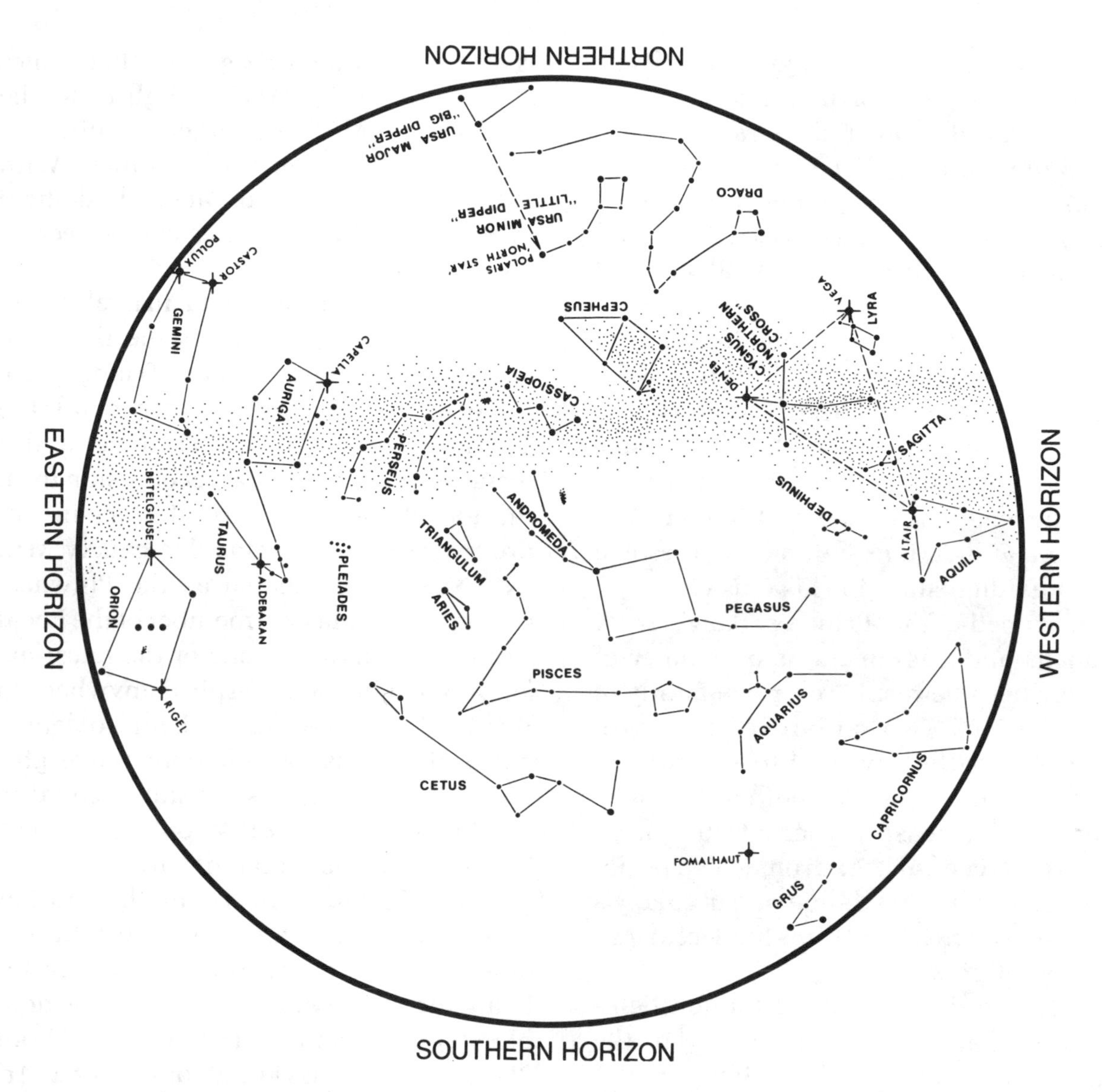

THE NIGHT SKY IN NOVEMBER

Latitude of chart is 34°N, but it is practical throughout the continental United States.

To use: Hold chart vertically and turn it so the direction you are facing shows at the bottom.

Chart time (Local Standard):
10 P.M. First of month
9 P.M. Middle of month
8 P.M. Last of month

Star Chart from GRIFFITH OBSERVER monthly magazine

scope shows a faint 9th magnitude companion just to the northwest.)

Two beautiful open clusters are easily found in Cassiopeia. The first, M52 (NGC 7654) lies about 6° to the northwest of β (beta) Cas—the westernmost star of the constellation. M52 contains about 100 stars, but binoculars are able to resolve about a dozen. The rest blur into a hazy cloud. The second cluster is M103 (NGC 581), just to the northeast of δ (delta) Cas—the second star from the left. This cluster is fairly compact and contains about 25 stars of magnitude 10.0 and fainter. Medium magnifications will begin to resolve the individual stars in this cluster.

Andromeda, the Princess daughter of Cepheus and Cassiopeia, is a member of the northern hemisphere's royal family of constellations. Two long arcs of stars mark her place in the sky, with α (alpha) Andromedae, or Alpheratz, both as her head and the northeast corner of the Great Square of Pegasus. Andromeda's special fame is that it is the home of our galaxy's twin—M31 (NGC 224), or the Great Andromeda Galaxy.

M31, at a distance of 2.2 million light-years, is the closest spiral galaxy to the Earth. Astronomers believe it resembles our own Milky Way galaxy. First noted by Persian astronomers as far back as the tenth century A.D., M31 is visible to the naked eye as a small, 3.5 magnitude fuzzy smudge about 5° northwest of μ (mu) And (the second star from the top of the upper arc of stars). In binoculars or a small telescope, a bright nucleus can be seen in the glow of the galaxy's extended halo. Again binoculars or a small telescope reveals two satellite galaxies of M31:M32 (NGC 221) and M110 (NGC 205). M32, just below M31, is a small, compact E2 galaxy that shines at magnitude 8.2; M110, slightly to the northwest of M31, is another satellite galaxy and has a low surface brightness. With moderate magnification and a wide-field eyepiece, all three galaxies can be seen in the same field of view.

To the east of Andromeda, in a relatively empty region of the sky, is the tiny constellation of Triangulum the Triangle. Located about halfway between α (alpha) Trianguli (the tip of the Triangle) and β (beta) Andromedae (the second star from the top of the lower arc of stars in Andromeda) is the spectacular galaxy M33 (NGC 598), also known as the Pinwheel Galaxy. This galaxy, a member of the Local Group of galaxies, is one of the finest examples of a face-on Sc spiral anywhere in the sky. It is sometimes difficult to detect, however, because of its low surface brightness. The object spans a distance equal to the diameter of the full Moon and can best be seen under low magnification.

Below Triangulum is another rather small triangle of stars of the constellation of Aries the Ram. Aries is best known for being the first sign of the zodiac. The double star λ (lambda) Ari is located about 2½° to the west of Hamal, or α (alpha) Ari (the northernmost star in the constellation). Moderate magnification will resolve the magnitude 4.9 and 7.7 white stars.

Stretching below Pegasus and Andromeda is the large V-shaped constellation of Pisces the Fishes. The Western Fish is marked by five faint stars in a group called the Circlet just below the Great Square of Pegasus. The Eastern Fish trails down to another watery constellation—Cetus, the Whale. There are no stars in Pisces brighter than magnitude 4.0. The star α (alpha) Pis-

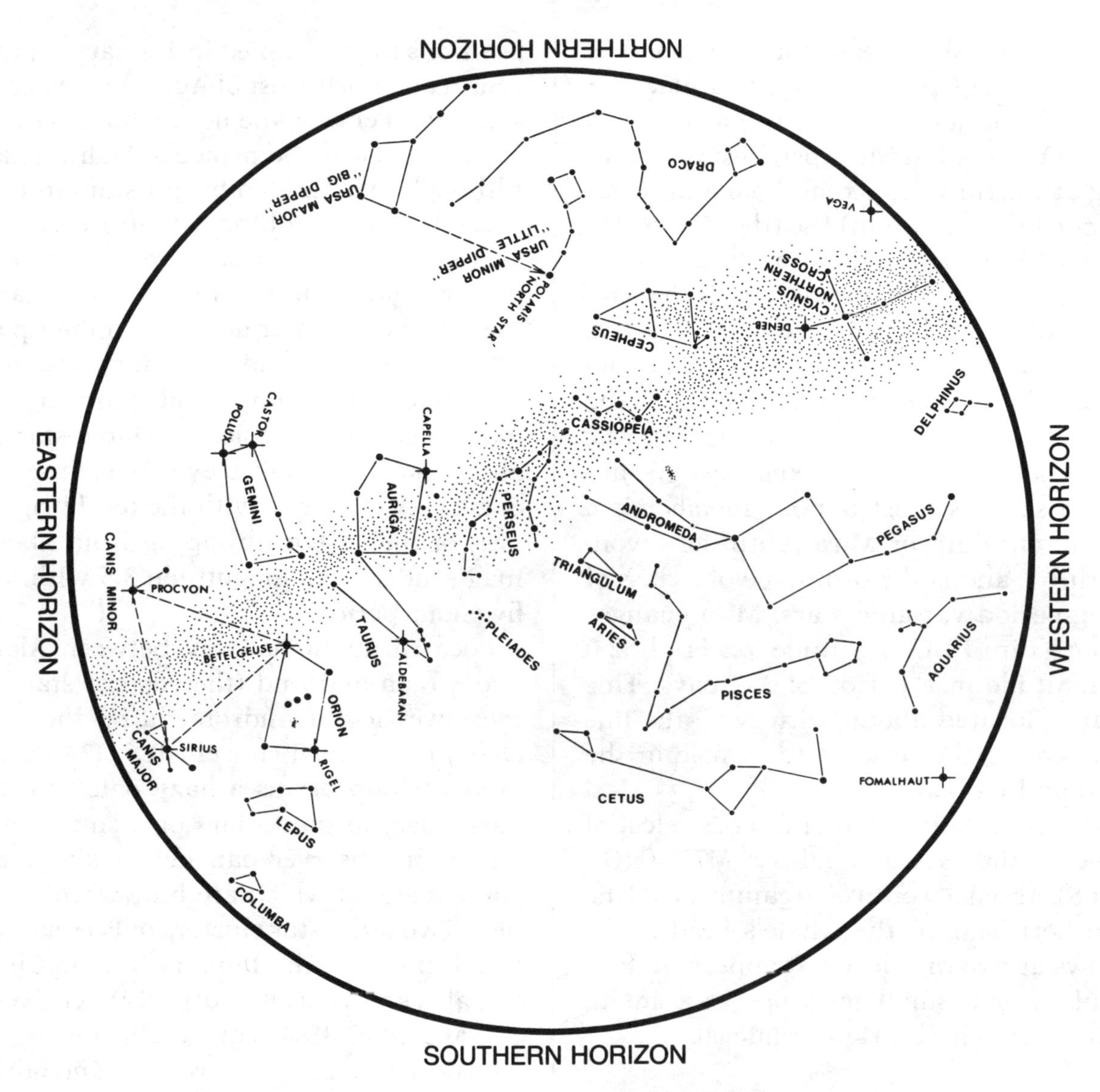

THE NIGHT SKY IN DECEMBER

Latitude of chart is 34°N, but it is practical throughout the continental United States.

To use: Hold chart vertically and turn it so the direction you are facing shows at the bottom.

Chart time (Local Standard):
10 P.M. First of month
9 P.M. Middle of month
8 P.M. Last of month

Star Chart from GRIFFITH OBSERVER monthly magazine

cium, or Alrisha (Arabic for "the knot"), is a double star. It shines at magnitudes 4.3 and 5.3 and joins the two fish at the apex of the V. In addition, ψ (psi) Piscium, shining at magnitudes 5.6 and 5.8 and about 2° northwest of χ (chi) Psc (the fourth star down from the northernmost star in the east), is also an interesting double that can be resolved with moderate magnification.

The final constellation crossing the meridian in the southern sky is Cetus the Whale. This constellation, like Pisces, stretches across a wide expanse of the autumn sky. Its most famous member is o (omicron) Ceti, or Mira (Latin for "wonderful"), and is the prototype of a class of long-period variable stars. Mira changes from a dim 10.1 magnitude to a bright 2.0 magnitude in a period of 332 days. This star is located about halfway on the line connecting the head of the whale in the east and its body.

Cetus contains another famous celestial object—the Seyfert galaxy M77 (NGC 1068). About 3° south of γ (gamma) Cet (the southern star in the whale's head), M77 glows at magnitude 8.9. It appears in binoculars or a small telescope as a small, faint disk with a brighter nucleus.

December

The year is ending and the Earth has almost completed one revolution around the Sun. The constellations are close to the positions they occupied in January. At the zenith and northwest of Auriga is the constellation Perseus, the hero who, in Greek legend, rescued the princess Andromeda. Although there are no bright stars in this constellation, Algol, or β (beta) Per, is a very good subject for observers interested in viewing variable stars. On the star chart, Algol is the fifth star down from the tip of the right branch of stars that form the constellation. An example of an eclipsing binary, Algol is the easiest variable star to observe with the naked eye. An eclipse occurs every 69 hours, with the total brightness of the system dropping from about magnitude 2.3 to magnitude 3.5 within a five hour period.

Located about halfway between Algol and γ (gamma) And (the eastern star on the lower leg of Andromeda on the star chart) is the open star cluster M34 (NGC 1039). It appears as a hazy patch to the naked eye; in binoculars or a small telescope, an observer can detect about 80 stars, many of which are blue-white doubles. Two other star clusters in Perseus are worth finding—the famous Double Cluster, also known as h and χ (chi) Per (NGC 869 and NGC 884). This double cluster is a favorite because they are close and fairly easy to observe. It is located about midway between the northernmost star in Perseus, η (eta) Per, and the easternmost star in Cassiopeia, ϵ (epsilon) Cass. A dark oval patch indicates its position on the star chart.

APPENDIX A

Basic Planetary Data

Name	Distance to Sun (millions of miles)	Period of Revolution	Period of Rotation [d = days, h = hours, m = minutes]	Equatorial diameter	Oblateness	Mass (Earth = 1)	Density (water = 1.0)	Surface Gravity (Earth = 1.0)	Albedo	Temperature	Inclination of Equator to Orbit	Inclination of Orbit to Ecliptic	Satellites
Mercury	min: 28.5 max: 43.5 avg: 36.4 (0.387 a.u.)	116 synodic days 88 sidereal days	58d 13h 30m	3,024 miles	0	0.055	5.43	0.38	11%	min: −300°F max: +700°	near 0°	7°	0
Venus	min: 66.5 max: 67.5 avg: 67.0 (0.723 a.u.)	584 synodic days 225 sidereal days	243d (retrograde)	7,504 miles	0	0.815	5.24	0.91	65%	cloudtops: 0°F surface: +900°F	177.3°	3.4°	0
Earth	avg: 93.0 (1.0 a.u.)	365.26 sidereal days	23h 56m	7,917 miles	0.0034	1	5.5	1	—	avg surface 80°F	23.45°	near 0°	1
Mars	min: 129 max: 154 avg: 142 (1.524 a.u.)	780 synodic days 687 sidereal days	24h 37m	4,208 miles	0.005	0.107	3.94	0.38	15%	at equator: min: −120°F max: +80°F	25.2°	1.8°	2
Jupiter	min: 460 max: 501 avg: 483 (5.203 a.u.)	399 synodic days 11.86 sidereal years	9h 55m	88,536 miles	0.0588	317.8	1.33	2.87	52%	cloudtops: −230°F	3.1°	1.3°	16
Saturn	min: 835 max: 935 avg: 885 (9.539 a.u.)	378 synodic days 29.46 sidereal years	10h 39m (varies with latitude)	70,400 miles	0.11	95.16	0.70	1.08	47%	cloudtops: −290°F	26.7°	2.5°	23 (at least)
Uranus	min: 1,700 max: 1,870 avg: 1,779 (19.182 a.u.)	369.66 synodic days 84.01 sidereal years	16h 48m	31,744 miles	0.03	14.50	1.30	0.91	51%	cloudtops: −370°F	97.9°	0.8°	15
Neptune	min: 2,770 max: 2,810 avg: 2,790 (30.058 a.u.)	367.49 synodic days 164.79 sidereal years	16h 3m	29,732 miles	0.026	17.20	1.76	1.19	41%	cloudtops: −390°F	29.6°	1.8°	8
Pluto	min: 2,750 max: 4,600 avg: 3,698 (39.785 a.u.)	367 synodic days 247 sidereal years	6d 9h 18m	1,395 miles	0	0.002	2.0 (approx.)	0.03 (?)	50% (?)	−390°F	?	17.1°	1

Table of Constellations

Name	Meaning	Date on Meridian
*Andromeda	Princess	Nov 10
Antlia	Air Pump	Apr 5
Apus	Bird of Paradise	Jun 30
*Aquarius	Water Bearer	Oct 10
*Aquila	Eagle	Aug 30
Ara	Altar	Jul 20
*Aries	Ram	Dec 10
*Auriga	Charioteer	Jan 30
*Boötes	Herdsman	Jun 15
Caelum	Chisel	Jan 15
Camelopardalis	Giraffe	Feb 1
*Cancer	Crab	Mar 15
Canes Venatici	Hunting Dogs	May 20
*Canis Major	Great Dog	Feb 15
*Canis Minor	Small Dog	Mar 1
*Capricornus	Sea Goat	Sep 20
Carina	Keel (of Argo)	Mar 15
*Cassiopeia	Queen	Nov 20
Centaurus	Centaur	May 20
*Cepheus	King	Oct 15
*Cetus	Whale	Nov 30
Chamaeleon	Chameleon	Apr 15
Circinus	Compass	Jun 15
*Columba	Dove	Jan 30
Coma Berenices	Berenice's Hair	May 15
Corona Australis	Southern Crown	Aug 15
*Corona Borealis	Northern Crown	Jun 30
*Corvus	Crow	May 10
Crater	Cup	Apr 25
Crux	Southern Cross	May 10
*Cygnus	Swan	Sep 10
*Delphinus	Dolphin	Sep 15
Dorado	Dorado	Jan 20

Name	Meaning	Date on Meridian
*Draco	Dragon	Jul 20
Equuleus	Colt	Sep 20
Eridanus	River	Jan 5
Fornax	Furnace	Dec 15
*Gemini	Twins	Feb 20
*Grus	Crane	Oct 10
*Hercules	Hero	Jul 25
Horologium	Clock	Dec 25
*Hydra	Water Serpent	Apr 20
Hydrus	Water Snake	Dec 10
Indus	Indian	Sep 25
Lacerta	Lizard	Oct 10
*Leo	Lion	Apr 10
Leo Minor	Small Lion	Apr 10
*Lepus	Hare	Jan 25
*Libra	Balance	Jun 20
Lupus	Wolf	Jun 20
Lynx	Lynx	Mar 5
*Lyra	Lyre	Aug 15
Mensa	Table	Jan 30
Microscopium	Microscope	Sep 20
Monoceros	Unicorn	Feb 20
Musca	Fly	May 10
Norma	Carpenter's Square	Jul 5
Octans	Octant	Sep 20
*Ophiuchus	Serpent Bearer	Jul 25
*Orion	Hunter	Jan 25
Pavo	Peacock	Aug 25
*Pegasus	Winged Horse	Oct 20
*Perseus	Hero	Dec 25
Phoenix	Phoenix	Nov 20
Pictor	Painter's Easel	Jan 20
*Pisces	Fishes	Nov 10
Piscis Austrinus	Southern Fish	Oct 10
Puppis	Poop (of Argo)	Feb 25
Pyxis	Ship's Compass	Mar 15
Reticulum	Net	Dec 30
*Sagitta	Arrow	Aug 30
*Sagittarius	Archer	Aug 20

Name	Meaning	Date on Meridian
*Scorpius	Scorpion	Jul 20
Sculptor	Sculptor	Nov 10
Scutum	Shield	Aug 15
*Serpens (Caput)	Serpent (Head)	Jun 30
*Serpens (Cauda)	Serpent (Tail)	Aug 5
Sextans	Sextant	Apr 5
*Taurus	Bull	Jan 15
Telescopium	Telescope	Aug 25
*Triangulum	Triangle	Dec 5
Triangulum Australe	Southern Triangle	Jul 5
Tucana	Toucan	Nov 5
*Ursa Major	Great Bear	Apr 20
*Ursa Minor	Little Bear	Jun 25
Vela	Sail (of Argo)	Mar 25
*Virgo	Virgin	May 25
Volans	Flying Fish	Mar 1
Vulpecula	Fox	Sep 10

* Constellations explored in Chapter 14.

APPENDIX C

Solar and Lunar Eclipses 1993–2010

TABLE 1 SOLAR ECLIPSES

Date		Type	Area of Visibility
1993	May 21	Partial	N America except SE, Arctic
	Nov 13	Partial	Antarctica
1994	May 10	Annular	NE Asia, N and C America, Arctic
	Nov 3	Total	C and S America, S Africa
1995	Apr 29	Annular	C and S America, Caribbean
	Oct 24	Total	Arabia, Asia, Japan, Australia, Pacific
1996	Apr 17	Partial	New Zealand, Pacific
	Oct 12	Partial	NE Canada, Greenland, Europe, N Africa
1997	Mar 8	Total	E Asia, Japan, NW N America
	Sep 1	Partial	S Australia, New Zealand, S Pacific
1998	Feb 26	Total	S and E N America, C America
	Aug 21	Annular	S and SE Asia, Indonesia, Australasia
1999	Feb 16	Annular	Indian Ocean, Antarctica, Indonesia
	Aug 11	Total	NE N America, Arctic, Europe, Arabia
2000	Feb 5	Partial	Antarctica
	Jul 1	Partial	S South America, SE Pacific Ocean
	Jul 31	Partial	NW North America
2001	Jun 21	Total	S Atlantic Ocean, S Africa
	Dec 14	Annular	Pacific Ocean, Costa Rica
2002	Jun 10	Annular	Celebes Sea to Mexico
	Dec 4	Total	S Africa to S Australia
2003	May 31	Annular	Iceland, Greenland
	Nov 23	Total	Antarctica
2004	Apr 19	Partial	S Africa, S Atlantic Ocean
	Oct 14	Partial	Japan, NE Asia, N Pacific Ocean
2005	Apr 8	Ann/Tot	SW Pacific Ocean, N South America
	Oct 3	Annular	Spain, N Africa, Kenya
2006	Mar 29	Total	W and N Africa, Central Asia
	Sep 22	Annular	Guyana, S Atlantic Ocean
2007	Mar 19	Partial	Asia
	Sep 11	Partial	S S America
2008	Aug 1	Total	Siberia, N China
2009	Jan 26	Annular	Indonesia, Borneo
	Jul 22	Total	India, S China, S Pacific Ocean
2010	Jul 11	Total	S Pacific Ocean

TABLE 2 LUNAR ECLIPSES

Date		Type
1993	Jun 4	Total
	Nov 29	Total
1994	May 24	Partial
1995	Apr 15	Partial
1996	Apr 3	Total
	Sep 26	Total
1997	Mar 23	Partial
	Sep 6	Total
1999	Jul 28	Partial
2000	Jan 21	Total
	Jul 16	Total
2003	May 16	Total
	Nov 7	Total
2004	May 4	Total
	Oct 28	Total
2007	Mar 3	Total
	Aug 28	Total
2008	Feb 21	Total
	Aug 16	Partial
2010	Jun 25	Partial

Kepler's Laws and the Universal Law of Gravitation

Kepler's First Law

The orbit of every planet is an ellipse, with the Sun as one focus of the ellipse. The other focus contains no object (Figure D. 1).

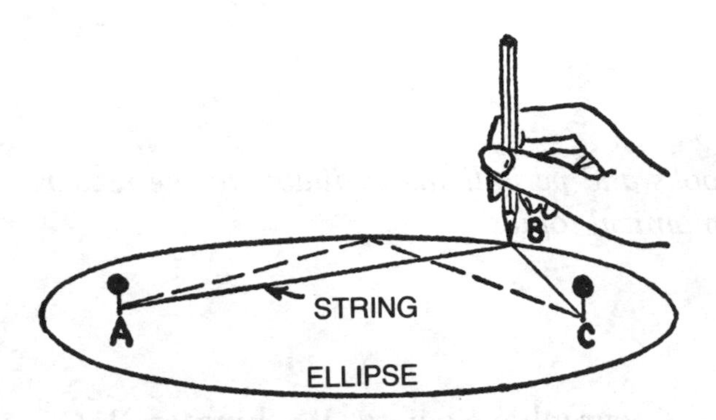

Figure D.1. Drawing of an ellipse. Fix the end of the string at points A and C. Stretch the string to form the angle B. Move the pencil about to form the oval curve. A and C are the foci of the ellipse.

Kepler's Second Law

A line from the Sun (known as a radius vector) to a planet sweeps across equal areas of space in equal intervals of time. This means that a planet moves faster as its orbit is nearer the Sun, perihelion, and slower as it moves farther away, aphelion (Figure D. 2).

Kepler's Third Law

The squares of the revolutionary periods (time) of any two planets are proportional to the cubes of their mean, or average, distances to the Sun. For any two planets designated A and B, this relationship can be stated in the following algebraic equation:

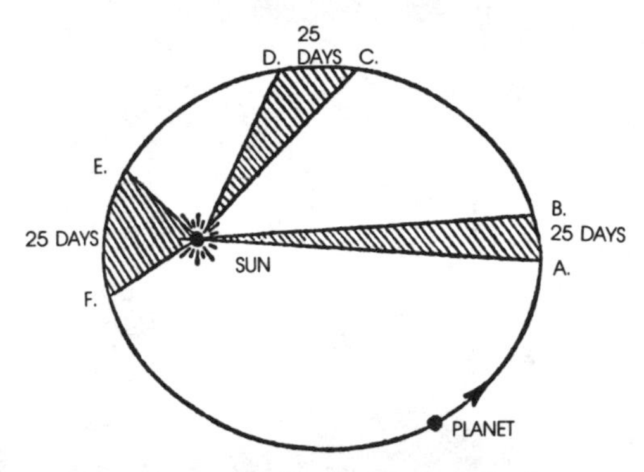

Figure D.2. The lines connecting the sun to A, B, etc., are radius vectors.

$$T^2 \text{ (of A)}/T^2 \text{ (of B)} = D^3 \text{ (of A)}/D^3 \text{ (of B)}$$

Newton's Universal Law of Gravitation

Every particle of matter in the universe attracts every other particle with a force that is proportional to the product of their masses, and inversely proportional to the square of the distance between them. This law can be expressed by the following algebraic equation:

$$F = G \times m_1 \times m_2/d^2$$

where F is the force, m_1 and m_2 are the masses, and d is the distance between them. G is the universal gravitational constant whose value is 6.7×10^{-8}, if m is expressed in grams, d in centimeters, and F in dynes.

For Further Reading

The following is only a sample of the many books and periodicals available to the reader interested in more detailed discussions of astronomical topics.

Introduction and Chapter 1

Friedman, H. *The Amazing Universe*, National Geographic Society, Washington, D.C., 1980.

———— *The Astonomer's Universe*, W.W. Norton, New York, 1990.

Jastrow, R. *Astronomy: Fundamentals and Frontiers* (4th ed.), John Wiley, New York, 1984.

Mallove, E. *The Quickening Universe*, St. Martin's Press, New York, 1987.

Pasachoff, J. *Astronomy: From the Earth to the Universe*, Saunders College Publishing, Orlando, FL, 1990.

Preston, R. *First Light*, New American Library, New York, 1987.

Chapter 2

Friedman, H. *Sun and Earth*, Freeman, New York, 1986.

Hartmann, W., ed. *Origin of the Moon*, Lunar and Planetary Institute, Houston, TX, 1986.

Meeus, J. *Canon of Solar Eclipses*, Pergamon Press, Elmsford, NY, 1966.

Moore, P. *New Guide to the Moon*, W.W. Norton, New York, 1976.

Chapter 3

Fraknoi, A., comp. *The Hubble Space Telescope*, Astronomical Society of the Pacific, 1991.

Goldsmith, D. *The Astronomers*, St. Martin's Press, New York, 1991.

Houk, R. *From the Hill: The Story of Lowell Observatory*, Lowell Observatory, Flagstaff, AZ, 1991.

Miczaika, G.R. *Tools of the Astronomer*, Harvard University Press, Cambridge, MA, 1961.
Page, T. *Telescopes*, Macmillan, New York, 1968.
Stott, C. *The Greenwich Guide to Astronomy in Action*, Cambridge University Press, New York, 1989.
Time-Life. *The New Astronomy*, Time-Life Books, Richmond, VA, 1989.
Woodbury, D.O. *The Glass Giant of Palomar*, Dodd, Mead, New York, 1953.

Chapters 4, 5, 6, and 7

Alexander, A.F. *The Planet Saturn*, Dover, London, 1980.
Beatty, J. *The New Solar System*, (3rd ed.), Cambridge University Press, New York, 1990.
Burgess, E. *Uranus and Neptune*, Columbia University Press, New York, 1988.
Cunningham, C. *Introduction to Asteroids*, Willmann-Bell, Richmond, VA, 1988.
Greeley, R. *Planetary Landscapes*, Allen & Unwin, New York, 1985.
Horowitz, N. *To Utopia and Back*, W.H. Freeman, New York, 1986.
Hunt, G. *Atlas of Uranus*, Cambridge University Press, New York, 1985.
_________ *The Planet Venus*, Faber & Faber, London, 1982.
Kelch, J. *Small Worlds*, Julian Messner, New York, 1991.
Littmann, M. *Planets Beyond: Discovering the Outer Solar System*, John Wiley, New York, 1989.
Malin, S. *The Greenwich Guide to the Planets*, Cambridge University Press, 1989.
Moore. P. *Mission to the Planets*, W.W. Norton, New York, 1990.
Morrison, D. *Voyages to Saturn*, NASA, Washington, DC, 1982.
NASA. *Atlas of Mercury* (NASA SP–423), Washington, DC, 1978.
NASA. *Geology of the Terrestrial Planets* (NASA SP–469), Washington, DC, 1984.
Sagan, C. *Comet*, Sky Publishing, Belmont, MA, 1989.
Schaaf, F. *The Starry Room*, John Wiley, New York, 1988.
Time-Life. *Comets, Asteroids, and Meteorites*, Time-Life Books, Richmond, VA, 1992.
_________ *The Far Planets*, Time-Life Books, Richmond, VA, 1988.
_________ *The Near Planets*, Time-Life Books, Richmond, VA, 1989.
Yeomans, D. *Comets: A Chronological History*, John Wiley, New York 1991.

Chapter 8

Mitton, S. *Daytime Star*, Scribner's, New York, 1981.
Time-Life. *The Sun*, Time-Life Books, Richmond, VA, 1990.
Wentzel, D. *The Restless Sun*, Sky Publishing, Belmont, MA, 1991.

Chapters 9 and 10

Asimov, I. *The Collapsing Universe: The Story of Black Holes*, Walker, New York, 1977.
Hartmann, W. *Cycles of Fire*, Workman, New York, 1987.

Kaufmann, W. *Stars and Nebulas*, W.H. Freeman, San Francisco, 1978.
Moore, P. *Astronomers' Stars*, W.W. Norton, New York, 1989.
Time-Life. *Stars*, Time-Life Books, Richmond, VA, 1988.

Chapters 11 and 12

Parker, B. *Colliding Galaxies*, Plenum Press, New York, 1990.
Time-Life. *Galaxies*, Time-Life Books, Richmond, VA, 1988.
Verschur, G. *Interstellar Matters*, Sky Publishing, Belmont, MA, 1990.

Chapter 13

Cohen, N. *Gravity's Lens: Views of the New Cosmology*, John Wiley, New York, 1988.
Parker, B. *Creation, The Story of the Origins of the Universe*, Plenum Press, New York, 1988.
———— *Invisible Matter and the Fate of the Universe*, Plenum Press, New York, 1989.

Chapter 14

Burnham, R. *Burnham's Celestial Handbook* (vols. 1–3), Willmann-Bell, Richmond, VA, 1987.
Chartrand, M. *Skyguide: A Field Guide to the Heavens*, Golden Press, New York, 1990.
Eicher, D. *The Universe From Your Backyard*, Cambridge University Press, New York, 1988.
Harrington, P. *Touring the Universe Through Binoculars*, John Wiley, New York, 1990.
Jones et al. (eds.). *Webb Society Deep-Sky Observers Handbook* (vols. 1–7), Willmann-Bell, Richmond, VA, 1975.
Karkoschka, E. *The Observer's Sky Atlas*, Springer-Verlag, New York, 1990.
Pasachoff, Jay M. *A Field Guide to the Stars and Planets*, Houghton Mifflin, New York, 1992.
Schaaf, F. *Seeing the Solar System*, John Wiley, New York, 1991.
Tirion, W. *Sky Atlas 2000.0*, Sky Publishing, Belmont, MA, 1981.
———— *Uranometria 2000.0* (vols. 1 and 2), Willmann-Bell, Richmond, VA, 1987.

Periodicals

Astronomy Monthly magazine with many articles on astronomy and observing; beautifully illustrated. Kalmbach Publishing Company, 21027 Crossroads Circle, P.O. Box 1612, Waukesha, WI, 53187.

Odyssey Monthly magazine designed to introduce children to the field of astronomy and space sciences. Cobblestone Press, Peterborough, New Hampshire.

Sky & Telescope Monthly magazine that covers astronomy and observing. Sky Publishing Corporation, Belmont, MA.

Glossary

aberration A flaw in the performance of any component in an optical system. Two examples are chromatic aberration and spherical aberration.

absolute magnitude The magnitude or brightness of a star if it were placed 10 parsecs from the observer.

absolute temperature A temperature scale (also known as the Kelvin scale) where the zero point represents the lowest attainable temperature possible. Absolute zero is nearly −459° F (−273° C).

accretion The process by which smaller particles of gas, dust, and rocks gradually join together to form larger bodies such as planetesimals, moons, and planets.

albedo The reflectivity of a celestial object, usually expressed as a percent.

alloy A physical (not chemical) combination of two or more metals.

altitude The distance above the horizon, usually expressed in degrees, of any celestial object.

angstrom An extremely small distance that is equal to 10^{-10} of a meter. It is used to measure the wavelength of light and other electro-magnetic radiation.

angular momentum The rate of spin of a rotating object.

aperture The diameter of a telescope's objective.

apogee The point in orbit at which any satellite is furthest from the Earth.

apparent magnitude A value that represents the amount of radiation (light) received by an observer from a celestial object without regard to the object's distance from the observer. The larger the value, the less radiation received.

arc minute A measure of angular distance that is equal to 1/60 of a degree.

arcsecond A measure of angular distance that is equal to 1/60 of an arc minute.

asterism A recognizable pattern formed by stars, usually within a constellation. For example, the Big Dipper is the asterism for Ursa Major.

astronomical unit The average distance between the Earth and the Sun, equal to 93 million miles.

aurora (plural is **aurorae**) A changing, diffused glow of light usually seen in the night sky at high latitudes in both hemispheres. They are caused by the interaction of the solar wind and charged particles in the Earth's atmosphere.

axis An imaginary line through the center of any rotating sphere.

azimuth The east-west direction along the terrestrial horizon in the altitude-azimuth (alt-az) system of celestial coordinates. See **altitude**.

big bang The cosmological theory that states the universe came into being in a gigantic explosion that created all matter, time, and space.

black dwarf The cold remains of a star whose nuclear fires have been extinguished.

black hole A collapsed star whose surface gravity is so great that nothing, not even light, can escape.

cannibal galaxy A massive galaxy that literally incorporates other galaxies within it.

carbon flash The violent explosion of a moderately massive star's carbon core that causes the entire star to shatter.

celestial equator A circle that represents the extension of the Earth's equator on the celestial sphere. See **celestial sphere**.

celestial sphere The imaginary sphere of the sky containing the stars and centered on the Earth.

centrifugal force An apparent force that pushes an object away from the center of rotation.

Chandrasekhar mass A star's highest mass value to be a white dwarf; it is equal to 1.4 times the mass of the Sun.

circumpolar stars Stars that never seem to set below the horizon, but instead revolve around the North Star.

configuration The arrangement of the components in any optical system.

conjunction The apparent line-up of the Sun, Earth, and a planet. Inferior conjunction is when an inferior planet is directly between the Earth and the Sun. Superior conjunction is when the planet is on the opposite side of the Sun from the Earth. See **opposition**.

cosmic background radiation The faint radiation in every direction of the sky; it appears to be the leftover energy from the big bang.

cosmology The study of the universe as a whole—its past, present, and future.

declination The angular distance of an object above or below the celestial equator, measured in degrees, minutes, and seconds. It is similar to divisions of latitude on Earth.

density The amount of matter in a given volume; usually used in reference to celestial objects.

dispersion The bending of light depending on its particular wavelength. For example, a prism is said to disperse light to create its colorful spectrum.

Doppler Effect The change in radiation wavelength due to the relative positions and velocities of a source and an observer.

eclipsing binary A double star system where one star periodically partially, or totally, eclipses the light from the other star.

ecliptic The great circle on the celestial sphere formed by the intersection of that sphere with the plane of the Earth's orbit. Or, the path described on the celestial sphere by the Sun during its apparent annual motion around the Earth.

electromagnetic radiation Waves produced by vibrating charged particles that travel through a vacuum at the speed of light. They range from long radio waves to short, high-energy gamma rays. See **spectrum**.

ellipse An oval shape (similar to a squashed circle) that describes the orbits of many planets.

elliptical galaxy A galaxy in the shape of an ellipse.

elongation Angular distance from the Sun, measured in degrees, minutes, and seconds of angle.

entropy The tendency toward disorder of any organized system.

equinox Two intersection points between the ecliptic and the celestial equator. When the Sun is at one of these two points, the length of day and night are equal everywhere on Earth; this occurs on or about March 21 (vernal equinox) and September 23 (autumnal equinox).

erg A very small unit of energy in the centimeter-gram-second system.

event horizon The boundary of a black hole where no particles of light can escape.

fixed stars A term applied by ancient observers to the seemingly unmoving stars on the celestial sphere. Used to distinguish them from the planets, comets, and meteors.

focal ratio (f/) On a telescope, the number derived by dividing the focal length by the aperture of the objective lens or primary mirror.

focal-length On a telescope, the distance between the center of a lens or mirror and the focus.

focus (plural is **foci**) The point on a lens where all the light rays parallel to the axis meet to form an image of an object.

G2 The spectral classification of our Sun.

galactic halo A spherical volume of matter surrounding a spiral galaxy whose radius is roughly equal to the galactic disk's radius. It contains numerous Population II stars and globular clusters.

galactic year The period of time it takes for the Sun to complete one revolution around the galaxy. It is equal to approximately 225 million years.

galaxy A large collection of stars, gas, and dust held together by mutual gravitation.

globular cluster A compact collection of approximately one million old stars found in a galactic halo.

gravitational lensing The gravitational bending of light by a very massive object such as a star or galaxy.

gravitational resonance The constant adjustment to an object's orbit such as the asteroids' orbital changes from Jupiter's gravity.

gravity The attraction of one mass to another; one of the basic forces in the universe.

greenhouse effect When a planet's atmosphere lets in visible solar radiation but blocks the escape of outgoing infrared radiation, thus creating a great increase in surface temperature.

helioseismology The science that monitors quakes and tremors within the Sun in order to understand its internal structure.

helium flash The explosive ignition of a helium-burning star's core.

Hertzsprung-Russell diagram A plot of the luminosity (or brightness) against the temperature (or spectral type) of a collection of stars. It is useful in showing a star's evolution.

Hubble constant The ratio of the velocity of recession to the distance of a galaxy. The ratio is between 50 and 100 kilometers per second per megaparsec.

Hubble's law The linear relationship between a celestial object's distance and red-shift due to the expansion of the universe.

inferior The planets inside the orbit of the Earth.

inflationary epoch The sudden and rapid expansion of the universe in the first milliseconds of creation.

inner planets The small, rocky planets that are relatively close to the Sun: Mercury, Venus, Earth, and Mars.

instability strip That portion of the Hertzsprung-Russell diagram that marks stars that evolved off the Main Sequence.

interferometry A method where two or more telescopes are used simultaneously to obtain a greater resolving power than a single telescope.

interstellar medium Matter found in the space between the stars.

ionized Refers to an atom stripped of one or more electrons.

irregular galaxy A galaxy that has no regular or geometric shape.

Jovian The large, gaseous planets that resemble (and include) Jupiter: Saturn, Uranus, and Neptune.

light-year The distance that light travels in one year; approximately 6×10^{12} miles.

local celestial meridian The imaginary circle that goes through the zenith and points north and south on the observer's horizon.

Local Group A local cluster of galaxies containing the Milky Way Galaxy.

magnitude A numerical value that indicates the luminosity of a celestial object. The larger the numerical value of the magnitude, the less the luminosity. See **absolute** and **apparent magnitude**.

Main Sequence A line on the Hertzsprung-Russell diagram representing the majority of stable stars. These stars derive their energy by the thermonuclear fusion of hydrogen into helium.

megaparsec A distance that represents one million parsecs, or 3.25 million light-years.

meteor A streak of light in the sky caused by the frictional heating of a piece of interplanetary material; also called a "shooting star."

meteor shower Meteors that appear to rain from a certain point in the sky over the course of several nights. Caused by the collision of the Earth's atmosphere with a swarm of interplanetary particles (usually the debris of a comet).

minor planet Another name for the numerous asteroids in the Solar System.

moon A natural satellite of a planet.

morning/evening star Not a star, but either the planet Mercury or Venus when seen in the morning or evening sky.

nebula A term given to any diffuse celestial object. Examples are planetary nebula, extragalactic nebula (galaxy), and emission nebula (a type of interstellar gas cloud).

neutrino A subatomic particle that has no mass or charge. It is believed to travel at the speed of light and can pass through large amounts of matter before being absorbed.

neutron star A star of extremely high density and composed mainly of neutrons; it is usually a star at the end of its life cycle.

nodes The intersection points of a celestial body's orbit with a plane.

nova A star that suddenly increases in brightness (by a factor of hundreds to thousands) then returns to normal; a nova can last for days or over a year.

nutation The wobble of a spinning object's rotational axis.

oblate sphere A sphere that is somewhat flattened at the poles.

occultation When a larger celestial object passes between a smaller object and an observer, thereby hiding the smaller object; for example, the Moon often occults a star.

opaque A material that blocks the passage of light.

open cluster Between 10 and 10,000 young stars found in a loose assemblage. These are found in the galaxy's disk and are also known as galactic clusters.

opposition When a superior planet and the Sun are on opposite sides of the Earth. To observers, the planet is rising above the eastern horizon just as the Sun is setting below the western horizon. See **conjunction**.

oscillating A variation of the big bang cosmology where the universe undergoes a cycle of repeated expansions and contractions.

outer planets The planets beyond the Asteroid Belt: Jupiter, Saturn, Uranus, Neptune, and Pluto.

parabolic/hyperbolic Two types of open-ended orbits followed by comets; they make one swing around the Sun and never return.

parallax The apparent shift in a celestial object's position due to a change in the observer's position. This method is used to determine the distance to stars.

parsec Equal to approximately 3.2 light-years.

penumbra The outer, lighter part of a shadow cast by a planet or satellite; or the gray outer region of a sunspot.

perigee The point in the orbit of an Earth-orbiting satellite closest to the Earth. See **apogee**.

period The interval of time it takes a satellite to complete one revolution around its parent planet.

perpendicular A line that intersects another line at a 90° angle.

photons Electromagnetic energy in its particle form (light).

planetary nebula A shell of glowing gas around a hot central star. The gas was originally ejected by an older star as it became a white dwarf.

planetesimals Small, planet-like objects that are believed to have formed from the original solar nebula.

plasma A gas that is extremely hot and electrically conductive.

polarity The property of light or radiant heat to arrange itself along magnetic lines.

Population I Stars with a wide range of ages and a high abundance of heavy elements. They are found mainly in galactic disks and spiral arms.

Population II Relatively old stars with a low abundance of heavy elements found in galactic halos and nuclei.

precession A slow motion of a rotating body's axis that causes the axis to trace out a cone in space.

primordial Original, generally refers to all of the elements in the universe at the time of the big bang.

prominence Arcs and loops of glowing gas that extend far above the surface of the Sun.

proper motion A star's angular velocity in a direction perpendicular to the line of sight of a terrestrial observer.

proto- A prefix that refers to a celestial body during its formative stages, for example, a protostar and protogalaxy.

protostar The part of a condensing nebula that is evolving toward thermonuclear ignition.

pulsar A rapidly rotating neutron star whose high-energy flashes of radiation sweep out into space; the energy is intercepted by terrestrial observers at regular intervals or pulses.

quasar Celestial objects with an energy output generally exceeding that of an entire galaxy. They show large red-shifts, making them the most distant observable objects in the universe.

radial velocity The velocity of a star in a terrestrial observer's line of sight.

radiant A point in the sky from which meteors seem to diverge.

radius One-half of a circle or sphere's diameter.

ram pressure stripping A process where the intergalactic medium resists the motion of a galaxy, stripping it of iron and other heavy elements.

reciprocal An inverse relationship between two mathematical values; for example, 2 is the reciprocal of ½.

red giant A large, luminous star with a cool surface. They have radii 15 to 30 times that of the Sun and are plotted in the upper right-hand corner of the Hertzsprung-Russell diagram.

red-shift The displacement of waves of electomagnetic radiation to longer wavelengths. A result of the expansion of the universe, high relative velocities, or strong gravitational fields.

refraction The bending of light when it passes from one medium to another, such as from air to water.

relativistic velocity Velocities approaching the speed of light.

resolution The degree to which fine details can be seen. In a telescope, the ability to separate two close celestial objects into two distinct units.

retrograde Apparent backward (westward) motion of a planet through a starfield; or the clockwise motion of a satellite around its parent planet.

rotational velocity The speed of an object's spin around its axis.

satellite A natural body in orbit around another body, usually a planet.

Seyfert galaxy A type of galaxy with an unusually bright, compact nucleus and faint spiral arms. They are a possible connecting link between normal galaxies and quasars.

sidereal day The time it takes the Earth to make one complete rotation on its axis with respect to the stars.

sidereal, synodic Two methods in determining the length of the month. The first uses the stars as a fixed reference point; the second uses the alignment of the Sun, Earth, and Moon in a straight line.

singularity The incredibly minute point at the center of a black hole where all the volume of the collapsed star has been compressed.

solar constant of radiation The quantity of radiant solar heat received by each square centimeter on the Earth's surface. It is equal to 1.94 calories per minute.

solar nebula The large cloud of gas and

dust that eventually formed the Sun and the other members of the Solar System.

solar wind The outward flow of particles such as protons and electrons from the Sun.

speckle interferometry Taking short photographic exposures of a star or other celestial object, then removing the effects of the Earth's atmosphere through computer analysis.

spectroscopic binary A two star system that can only be detected with the use of a spectroscope.

spectrum The pattern of colors or radiation intensities at different wavelengths presented in order of their wavelengths; or the appearance of light after it has been dispersed by a prism or a grating.

spinar The name for a theoretically spinning black hole.

spiral galaxy A galaxy that resembles a huge rotating pinwheel with a dense, bright nucleus and luminous arms trailing into space.

steady-state A cosmological theory where the universe has no beginning or end but is constantly renewing itself.

superior Refers to the planets outside the orbit of the Earth.

supernova A massive star that explodes with a hundred-billionfold increase in luminosity. Believed to be caused by the core's sudden detonation or its rebound from a rapid collapse.

sunspot A dark, relatively cool spot visible on the Sun's surface; associated with strong magnetic fields.

synodic (see **sidereal**)

tangential velocity The velocity of a star across an observer's line of sight.

terrestrial Refers to the planets that are relatively small with solid, rocky surfaces: Mercury, Venus, Earth, Mars, and perhaps Pluto.

thermonuclear fusion The process where hydrogen fuses into the heavier helium, releasing tremendous amounts of energy; the process by which stars shine.

transit When an inferior planet (Venus or Mercury) crosses the face of the Sun.

translucent Material that allows the passage of light but not clearly.

turbulence A theory that states that the planets of our Solar System were formed in cells, where faster moving gas and dust met slower moving gas and dust.

umbra The darkest part of a shadow cast by a celestial body or the darkest portion of a sunspot.

variable star A star whose luminosity or spectrum appears to change with time.

velocity-distance relationship The more distant the galaxy, the faster it is receding from us; also known as **Hubble's Law**.

visual binary A double star system where the stars can be seen with a telescope or the naked eye.

voids Large, dark areas in the universe that appear to contain no galaxies or matter.

wave density A theory that the spiral arms of a galaxy are maintained by local densities of gas and dust that lead to star formation.

wavelength The distance between two successive crests or valleys in a wave; determines the type of electromagnetic radiation.

white dwarf The final stage in the evolution of solar-like Main Sequence stars about the size of the Earth.

zenith The point on the celestial sphere directly overhead.

zodiac A band on the celestial sphere, 8° wide on each side of the ecliptic. It is divided into 12 equal sections, each identified by a constellation.

zone Usually refers to the bright belts that are visible at the top of Jupiter's atmosphere.

zone of avoidance The fewer visible galaxies near the galactic plane because of interstellar material concentrations that block the galaxies in that direction.

INDEX